Optimal Strategies in Sports Economics and Management

Sergiy Butenko · Jaime Gil-Lafuente ·
Panos M. Pardalos
Editors

Optimal Strategies in Sports Economics and Management

Foreword by Jaime Gil-Aluja

Springer

Editors
Dr. Sergiy Butenko
Texas A & M University
Department of Industrial and
Systems Engineering
236E Zachry Engineering Center
College Station, TX 77843-3131
USA
Butenko@tamu.edu

Prof. Dr. Jaime Gil-Lafuente
University of Barcelona
Department of Economics
Av. Diagonal 690
08034 Barcelona
Spain
j-gil@ub.edu

Prof. Panos M. Pardalos
University of Florida
Department of Industrial and
Systems Engineering
303 Weil Hall
P. O. Box 116595
Gainesville, FL 32611-6595
USA
pardalos@ufl.edu

ISBN 978-3-642-13204-9 e-ISBN 978-3-642-13205-6
DOI 10.1007/978-3-642-13205-6
Springer Heidelberg Dordrecht London New York

Library of Congress Control Number: 2010930921

Cover design: WMXDesign GmbH, Heidelberg

Printed on acid-free paper

Springer is part of Springer Science+Business Media (www.springer.com)

To our families

O mother of gold-crowned contests, Olympia, queen of truth; where men that are diviners observing burnt-offerings make trial of Zeus the wielder of white lightnings, whether he hath any word concerning men who seek in their hearts to attain unto great prowess and a breathing-space from toil; for it is given in answer to the reverent prayers of men–do thou, O tree-clad precinct of Pisa by Alpheos, receive this triumph and the carrying of the crown.

Pindar (522 BC – 443 BC)
Greek lyric poet

Foreword

During the last century, we have witnessed the birth and evolution of sport as an economic activity, which has created jobs on the one hand, but also problems of management on the other. This process has not been immune from the particular characteristics associated with sport, typically united here more than in other activities: technique, physical effort, entertainment and passion. And all this within a framework of ever-increasing consumption of financial resources. It is not surprising, therefore, that commonly-used economic models, based on mechanistic approaches, do not provide a viable solution to increasingly complex and increasingly frequent problems. Any attempt to apply such an approach in this technical, economic and financial context can only result in failure. The high degree of subjectivity inherent in sporting activity requires new tools, in which remodeled conceptual, theoretical and technical elements should play an important role. Complexity, uncertainty and subjectivity are therefore basic to understand, and deal with, the phenomenon of sport.

The necessity of resorting to these elements was identified over a quarter of a century ago by a small group of professors and researchers at the University of Barcelona. Together we started the first postgraduate courses and organized seminars to alert sports centre managers, as well as to make private and public organizations aware of the increasing importance of a proper, specific management for sports organizations. For that reason we created the first course in "Economic Management for Sports Organizations" in 1991. Some years later, in 1995, Professor Ana María Gil-Lafuente started the course "Law, Finance and Taxation in Sports Organizations". Later, in 2000, Professor Jaime Gil-Lafuente, from the same research group, led a new course on "Strategic Marketing Management in Sports Organizations". These three courses are still given today as university extension courses. These teaching activities were made possible at that time thanks to the decisive support of the then Barcelona Football Club president, José Luís Núñez. His long-term vision enabled the publication of two works which opened the way for cutting-edge techniques for the analysis and management of sports organizations and activities. The first work, "The Universities in the Centenary of Football Club Barcelona", consists of a series of works written by researchers from Catalan universities, which were collected together in a book published in 1999, in the context of studies in the field of sport. The second is, in our opinion, a more advanced piece of research, using techniques related to uncertainty. In this work, multivalent logics are applied for the first time ever to the study of uncertain phenomena inherent in the practice and management of sport. We are referring here to the book written by Professor Jaime Gil-Lafuente, "Algorithms for Excellence. Keys to success in sports management". This teaching and research activity has been accompanied by articles in important journals, presentations, seminars and discussions.

The Royal Academy of Economic and Financial Sciences of Spain, always open to proposals of collaboration containing new ideas in economic research, could not ignore a call to support an initiative to hold a meeting in its headquarters, organized by a Spanish group headed by Dr. Jaime Gil-Lafuente and a group from the University of Florida, led by Dr. Panos Pardalos, with an important and active participation of Dr. Sergiy Butenko of Texas A&M University. The purpose of the meeting was to exchange suitable ideas, concepts, methods and techniques to apply to the complex problems generated by both amateur and professional sport, or, to put it in another way, active sport and sport as entertainment. The result has been the book we now have the pleasure of presenting. The content of the work is mainly based on the conference titled "Economics, Management and Optimization in Sports After the Impact of the Financial Crisis" (EMOS), together with the achievements of the constant research activities carried out by the The Royal Academy of Economic and Financial Sciences of Spain for more than two and a half centuries in response to its primary vocation to serve the society. The first handful of researchers, of whom we may consider ourselves to be descendants and followers, began the first scientific tasks with ideals which have survived intact to the beginning of the new millenium.

We would also like to recognize the collaboration of the academician Llorenç Gascón, Vice-President of the Royal Academy of Economic and Financial Sciences, and the sports critic Josep Pons, for their excellent work in the leading and coordination of discussion groups that highlighted the conference. The different nature of these discussions, the first having scientific content, and the second centered on the human and social context, added to the diversity and enriched the contributions of the two research groups.

However, above all, the Royal Academy of Economic and Financial Sciences of Spain would like to express its acknowledgement to the authors of the chapters that contributed to this work. They have facilitated the multi-faceted vision of a social context, sport seen from different points of view and from different countries. From France, Professors Lionel Maltese and Lucien Veron; from Canada, Professors Brad R. Humphrey and Daniel S. Mason; from Italy, Professors Francesco Carlo Morabito and Domenico Marino; from Sweden, Mr. Patrick Siegbahn; from England, Professors Rob Simmons and Stefan Szymanski; from the USA, Professors Panos M. Pardalos, Qijpeng P. Zheng, Yingyan Lou and Donald Hearn and finally, from Spain, Professors Ana María Gil-Lafuente, Jaime Gil-Lafuente and Jaime Gil-Aluja. To all of them, our Royal Corporation wishes to express its gratitude for their efforts and also to congratulate them on their brilliant and unselfish work.

Barcelona, Spain Jaime Gil-Aluja

Preface

This volume comprises a collection of papers, most of which are based on selected presentations at the international conference "Economics, Management and Optimization in Sports. After the Impact of the Financial Crisis" that took place in Barcelona, Spain, on December 1–3, 2009, enriched by several additional invited contributions from distinguished researchers around the world. The topics covered by the papers in this collection include

- strategies for player selection in team sports;
- a framework for formal description of game systems;
- analysis of the impact of the financial crisis on professional sports in North America;
- resource allocation strategies in professional sports;
- top European football clubs after the crisis;
- pros and cons of building a new stadium in an uncertain environment;
- use of complex networks in sports applications;
- examining fairness in fourball golf competition;
- study of the impact of revenue sharing in English football;
- analysis of the relation of the financial structure of football clubs to the economic cycle;
- gambling strategies based on stochastic optimization techniques.

We would like to thank The Royal Academy of Economic and Financial Sciences of Spain for organizing the memorable conference in Barcelona; the authors of the chapters for their dedication to this project that resulted in this excellent collection; the anonymous referees for their timely and constructive reports; and Springer management and staff for the support and technical assistance.

College Station, TX — Sergiy Butenko
Barcelona, Spain — Jaime Gil-Lafuente
Gainesville, FL — Panos M. Pardalos

Contents

Contributors

Jaime Gil-Aluja University of Barcelona, 08034 Barcelona, Spain, jaigil@far.ub.es

Anna M. Gil-Lafuente University of Barcelona, 08034 Barcelona, Spain, amgil@ub.edu

Jaime Gil-Lafuente Department of Business Administration, University of Barcelona, 08034 Barcelona, Spain, j.gil@ub.edu

Donald Hearn University of Florida, Gainesville, FL, USA, hearn@ise.ufl.edu

Brad R. Humphreys Department of Economics, University of Alberta, Edmonton, AB, Canada T6G 2H4, brad.humphreys@ualberta.ca

Yingyan Lou Department of Civil, Construction and Environmental Engineering, The University of Alabama, Tuscaloosa, AL 35487, USA, ylou@eng.ua.edu

Lionel Maltese Paul Cezanne University, Aix-Marseille, France; Euromed Management, Marseille, France, lionel.maltese@univ-cezanne.fr

Domenico Marino Mediterranea University of Reggio Calabria, Reggio Di Calabria, Italy, dmarino@unirc.it

Daniel S. Mason Faculty of Physical Education & Recreation, University of Alberta, Edmonton, AB, Canada T6G 2H9, dmason@ualberta.ca

Francesco Carlo Morabito University Mediterranea of Reggio Calabria – DIMET, Reggio Di Calabria, Italy, morabito@unirc.it

Panos M. Pardalos Center for Applied Optimization, Department of Industrial & Systems Engineering, University of Florida, Gainesville, FL 32611, USA, pardalos@ufl.edu

Patrick Siegbahn Royal Institute of Technology, Stockholm, Sweden, patrick@siegbahn.com

Rob Simmons Lancaster University, Lancaster, UK, r.simmons@lancaster.ac.uk

Stefan Szymanski Cass Business School, City University London, 106 Bunhill Row, London EC1Y 8TZ, UK, szy@city.ac.uk

Lucien Veran Paul Cezanne University, Aix-Marseille, France; Euromed Management, Marseille, France, veran.lucien@wanadoo.fr

Ian Walker Lancaster University, Lancaster, UK, ian.walker@lancaster.ac.uk

Qipeng P. Zheng Center for Applied Optimization, Department of Industrial & Systems Engineering, University of Florida, Gainesville, FL 32611, USA, zqp@ufl.edu

Affinity in the Selection of a Player

Jaime Gil-Aluja and Anna M. Gil-Lafuente

Abstract In recent research on sports management, Gil-Lafuente (2002) puts forward a set of algorithms that are capable of resolving, from a scientific point of view, the problems faced by every person responsible for allocation of financial resources of a sports club. One of the questions of interest is the formation of groups of players who are substitutable among each other within the framework of team play. This is an essential matter that should be addressed prior to obtaining an order of preference among them. The proposed algorithms that are capable of providing a good solution to the problem were based on the theory of affinities. Inspired by these results, we propose an extension of the algorithm for the selection of a player, contributing elements that allow us to arrive at more general results. With this, and with no modification whatsoever, we have opened up a new path in the treatment of the proposed problem, in the event of a certain amount of uncertainty in the information. To address this case, we have started out with estimates made by means of intervals and also triplets or quadruples of confidence. Finally we have presented, within the extended concept of affinity, an algorithm based on the product of relations. We conclude by pointing out some alternative approaches to the concept of affinity, which we plan to develop in future works.

1 Similarities and Affinities

The notions of similarity and affinity represent different ways of expressing the concept of neighborliness. Similarity indicates a specific resemblance, either partial or total, between two physical or imaginary objects, in our case players. Affinity refers to a collective behavior of objects with respect to certain specific criteria, even when these criteria are not particularly well specified. But both of these notions indicate whether two or more objects (players), are related under adequate conditions with respect to explicit criteria for affinity.

J. Gil-Aluja (✉)
University of Barcelona, 08034 Barcelona, Spain
e-mail: jaigil@far.ub.es

S. Butenko et al. (eds.), *Optimal Strategies in Sports Economics and Management*,
DOI 10.1007/978-3-642-13205-6_1, © Springer-Verlag Berlin Heidelberg 2010

Similarity is based in the notion of distance. It is well known that distance can be defined in many different ways. From the mathematical point of view, the definition of distance that is best adapted to the problem under treatment is selected. The smaller is the distance between two objects, the greater is their similarity. Our objective consists in defining a relationship of similarity between n different objects (players) in order to be able to determine those objects that are similar at a given level α of similarity, by means of maximum sub-relations. This is done in the same manner as for affinities, however, in this case the maximum sub-relations are relations of resemblance, that is to say they are transitive. The word transitive at this juncture has a great importance since it is essential to distinguish between similarity and resemblance.

It is for this reason that, unlike with similarity, the notion of resemblance introduces the notion of *disjointed classes*. There are many reasons, which make it important to carry out the decomposition of a relation of similarity into maximum sub-relations of resemblance. In the particular case of a relation of similarity there exists an algorithm known as the Pichat algorithm (see, e.g., (Gil-Lafuente, 2002)).

Let us recall the *algorithm of maximum inverse correspondence* developed by Gil-Lafuente (2002) for finding affinities in sports applications. It consists of the following steps. Let E_1 represent the set of players and E_2 represent the set of criteria that a player may be evaluated on.

1. Select among E_1 and E_2 the set with a smaller number of elements. Without loss of generality, assume that this is set E_1.
2. For the smaller set (E_1), its *power set* is constructed, that is to say, the set of all its subsets.
3. Make every set-element of the *power set* to correspond to the set of criteria that are satisfied by each member of this set-element. This is the so-called "*connection to the right*".
4. For every set of criteria that is not void of the connection to the right, select the corresponding set of the power set with the greatest number of elements.
5. The relations between the sets that are arrived at form a Galois lattice, which, apart from showing the different homogeneous groupings, allows for the perfect structuring of the same.

The use of the notion of distance and the concept of affinity yield different results. By means of the distances we arrive at the group of players such that the players within the group are closer to one another. On the other hand, by means of affinities the players are grouped according to the selected criteria. We are therefore dealing with different notions. One of these allows us obtain an approximation of resemblance mathematically speaking, while with the other we arrive at affinities relative to certain criteria.

2 Affinities with Imprecise Preferences

In this section, we aim to extend the work of Gil-Lafuente (2002) described above. Given the very high level of subjectivity that exists in the assignment of values to the criteria by means of a number $x \in [0, 1]$ in the study on player selection, we now propose a model that consists in starting out from information expressed by means of confidence intervals (segment) $[x_1, x_2] \subset [0, 1]$. In this case the algorithm can still be applied for seeking affinities, but taking into account a very slight variation.

To explain this, we will use an example, in which the group of players is given by

$$E_1 = \{a, b, c, d\}$$

and the set if criteria consists of the following elements:

$$E_2 = \{A, B, C, D, E, F\}.$$

Let us assume that after consulting with the corresponding experts we have the following matrix:

	A	B	C	D	E	F
a	0.8	[0.3, 0.5]	[0.0, 0.2]	[0.6, 0.9]	0.8	[0.3, 0.5]
b	0.2	[0.5, 0.6]	[0.5, 0.8]	1	0.6	[0.9, 1]
c	[0.3, 0.5]	0.5	[0.7, 0.9]	[0.8, 1]	[0.5, 0.6]	0.9
d	0	[0.3, 0.5]	[0.6, 0.8]	0.7	[0.3, 0.7]	[0.8, 1]

We will apply the algorithm of affinities of the maximum inverse correspondence with each number $x \in [x_1, x_2]$ defined by the introduced level α. For example, let us take $\alpha = 0.7$. Then we obtain the following matrix:

	A	B	C	D	E	F
a	1			1	1	
b			1	1		1
c			1	1		1
d			1	1	1	1

$$\begin{array}{rlrl} {}^*a & \longrightarrow ADE & abc & \longrightarrow D \\ b & \longrightarrow CDF & abd & \longrightarrow D \\ c & \longrightarrow CDF & acd & \longrightarrow D \\ {}^*d & \longrightarrow CDEF & {}^*bcd & \longrightarrow CDF \\ ab & \longrightarrow D & {}^*abcd & \longrightarrow D \\ ac & \longrightarrow D & & \\ {}^*ad & \longrightarrow DE & & \\ bc & \longrightarrow CDF & & \\ bd & \longrightarrow CDF & & \\ cd & \longrightarrow CDF & & \end{array}$$

resulting in:

$$\begin{array}{rl} a & \longrightarrow ADE \\ d & \longrightarrow CDEF \\ ad & \longrightarrow DE \\ bcd & \longrightarrow CDF \\ abcd & \longrightarrow D \end{array}$$

Therefore:

	A	D	E
a	0.8	[0.7, 0.9]	0.8

	C	D	E	F
d	[0.7, 0.8]	0.7	0.7	[0.8, 1]

	D	E
a	[0.7, 0.9]	0.8
d	0.7	0.7

	C	D	F
b	[0.7, 0.8]	1	[0.9, 1]
c	[0.7, 0.9]	[0.8, 1]	0.9
d	[0.7, 0.8]	0.7	[0.8, 1]

	D
a	[0.7, 0.9]
b	1
c	[0.8, 1]
d	0.7

The corresponding Galois lattice is shown below.

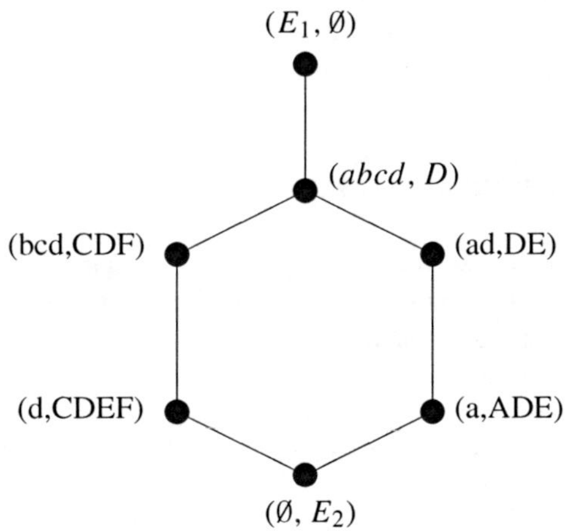

In this lattice, which is assumed to be constructed based on information provided by experts of a sporting institution, it can be seen that as we move from bottom to top, the groups of players increase in size, while the number of criteria satisfied by each player in the group decreases. This way we have a compact representation of various possible combinations of players that will facilitate the process of selecting the group that satisfies the criteria considered as most important or essential.

3 Affinity Indexes

At this point, the following question can be raised: how to evaluate affinity by means of an index that refers to all the sub-relations? We propose to use the average of the corresponding valuations of the players defining the sub-relation with respect to the criteria involved in the sub-relation. Let x_{ij} denote the average of the interval defining the valuation of player i with respect to the criterion j. Then for a pair (I, J) defining a sub-relation for the set of players I and the set of criteria J we compute

$$p(I, J) = \frac{1}{|I||J|} \sum_{i \in I} \sum_{j \in J} x_{ij} \tag{1}$$

For instance, for the level $\alpha = 0.7$ in our example we have:

$$
\begin{aligned}
p_{0.7}(a, ADE) &= \tfrac{1}{3}(0.8 + 0.8 + 0.8) = 0.800 \\
p_{0.7}(d, CDEF) &= \tfrac{1}{4}(0.75 + 0.7 + 0.7 + 0.9) = 0.762 \\
p_{0.7}(ad, DE) &= \tfrac{1}{4}(0.8 + 0.8 + 0.7 + 0.7) = 0.750 \\
p_{0.7}(bcd, CDF) &= \tfrac{1}{9}(0.75 + 1 + 0.95 + 0.8 + 0.9 + 0.9 + 0.75 + 0.7 + 0.9) \\
&= 0.850 \\
p_{0.7}(abcd, D) &= \tfrac{1}{4}(0.8 + 1 + 0.9 + 0.7) = 0.850
\end{aligned}
$$

Obviously, $p \geq \alpha$, since α is the lower bound on valuation. The index of affinity for each sub-relation is naturally a very significant element for the considered problem.

4 Affinities with Confidence Triplets

Given the importance of the selection of a player, with transfer fees often evaluated in millions of dollars, it is essential to make the best possible use of the available information in making the selection decisions. To fine tune the estimates of information, which are charged with a very high degree of uncertainty, we propose placing a new value (or interval, if required) between the extremes of each given interval, which expresses our best estimate of the corresponding value. This way we obtain confidence triplets. Let us recall that a confidence triplet is presented, in its more general form, as follows:

$$a = (a_1, [a_2, a_3], a_4), \;\; a_1 \leq a_2 \leq a_3 \leq a_4 \in [0, 1].$$

If $a_2 < a_3$ then we have a confidence quadruple. In particular, the case where $a_2 = a_3$ is written as

$$a = (a_1, a_2, a_3), \;\; a_1 \leq a_2 \leq a_3 \in [0, 1].$$

Then $[a_2, a_3]$ in the confidence quadruple and a_2 in the confidence triple are called *the maximum of assumption*.

Frequently such representations are preferred since they allow to address the numerical subjectivity of a human operator faced with the ever-present uncertainty in the sporting sphere. Indeed, confidence triplets provide the human operator more flexibility than the confidence intervals, while still not requiring the precise measurements.

If we are dealing with the problem of obtaining affinities starting based on the given triplets, the methodology will be similar to the case when the available valuations were represented using values or intervals in [0, 1] as we have seen above.

Let us recall that for a quadruple $(a_1, [a_2, a_3], a_4)$ the number a that is most suitable for representing it could be the average of the four numbers defining the quadruple:

$$a = (a_1 + a_2 + a_3 + a_4)/4.$$

Thus, for the triple (a_1, a_2, a_3) we have:

$$a = (a_1 + 2a_2 + a_3)/4.$$

For the notion of affinity we will require the normalized distance to 1 (perfect affinity). We arrive at

$$d((a_1, [a_2, a_3], a_4), 1) = 1 - (a_1 + a_2 + a_3 + a_4)/4$$

in the first case and at

$$d((a_1, a_2, a_3), 1) = 1 - (a_1 + 2a_2 + a_3)/4$$

in the second case. In fact, it can be stated that for those relations in which the valuations are expressed by means of confidence triplets, in order to arrive at affinities we will have to operate in the same manner as for the valuations in numbers or in intervals in [0, 1]. We now present an example. Consider the following matrix:

	A	B	C	D
a	(0.2, [0.3, 0.4], 0.7)	[0.7, 0.8]	0.6	(0.5, [0.5, 0.6], 0.9)
b	(0.1, [0.2, 0.3], 0.5)	0.9	1	(0, [0.1, 0.2], 0.4)
c	0	[0.6, 0.8]	(0.8[0.8, 0.9], 1)	[0.7, 1]

Below we look for the affinities at a certain level, for example $\alpha = 0.7$.

	A	B	C	D
a	1	1		1
b		1	1	
c	0	1	1	1

Thus, we obtain:

	A	B	C
a	0.7	[0.7, 0.8]	[0.7, 0.9]

	B	C	D
c	[0.7, 0.8]	(0.8, [0.8, 0.9], 1)	[0.7, 1]

	B	D
a	[0.7, 0.8]	[0.7, 0.9]
c	[0.7, 0.8]	[0.7, 1]

	B	C
b	0.9	1
c	[0.7, 0.8]	(0.8, [0.8, 0.9], 1)

	B
a	[0.7, 0.8]
b	0.9
c	[0.7, 0.8]

Now we can calculate the relative indices of affinity at any level. For that we use formula (1) with the level $\alpha = 0.7$. It can be seen, once again, that the indices of affinity are always greater than or equal to the respective levels α. The construction of the corresponding Galois lattices is then straightforward.

5 Affinities and Groupings

In the problems of player selection, affinities can be found based on the qualifications (properties) of players with respect to the given selection criteria. Based on these properties, we want to find the players who have the same or better qualifications with respect to the criteria used. Some of them can be admitted or rejected starting out from sets or sub-sets of properties. All those possessing the same properties form groupings that can be considered as affinities. Their aptitudes and attitudes join them together by affinities. It is for this reason that below we are going to introduce the selection of players with a certain amount of detail. As is well known arriving at affinities can be done using different procedures, some more effective than others, according to the nature of the situation that is being treated. Below we are presenting an algorithm, in which recourse is made to a product of relations.

We consider the first relation:

$$R_1 = \text{Player} \times \text{Capacities}$$

and the second relation

$$R_2 = \text{Capacities} \times \text{Position on the team.}$$

The composition $R_3 = R_1 \circ R_2$ ($\circ = \max\min$) will provide us with the positions on the team for each player. It should be considered that the sets R_1, R_2 and R_3 that are considered in practice are usually very large, so that they are broken down into smaller sections for data processing.

Let us consider the following example.

$E_1 = \{a, b, c, d, e\}$	5 players
$E_2 = \{\alpha, \beta, \gamma, \delta, \epsilon, \rho\}$	6 capabilities or aptitudes
$E_3 = \{A, B, C, D\}$	4 positions on the team

Let the two relations of interest be

$$\underset{\sim}{R} \subset E_1 \times E_2 \text{ and } \underset{\sim}{S} = E_2 \times E_3.$$

The following matrices provide the estimates of numerical values describing the considered relations:

$\underline{R} =$

	α	β	γ	δ	ε	ρ
a	0.3		0.8	1	0.5	
b	0.7	0.2	0.4		1	0.8
c	0.1			0.3		1
a	0.2	0.5	0.5	0.4		0.6
a	1	1	0.2		0.1	0.7

$\underline{S} =$

	A	B	C	D
α		1	0.6	
β	0.8	0.7		0.2
γ	1	0.4		0.3
δ		0.8		0.8
ε	0.5	1	1	
ρ		0.2	0.6	0.7

The data describing $\underline{R}$ and $\underline{S}$ were obtained through conversations, questions, surveys, measurements or other valuations. We have:

	α	β	γ	δ	ε	ρ
a	0.3		0.8	1	0.5	
b	0.7	0.2	0.4		1	0.8
c	0.1			0.3		1
a	0.2	0.5	0.5	0.4		0.6
a	1	1	0.2		0.1	0.7

$\circ$

	A	B	C	D
α		1	0.6	
β	0.8	0.7		0.2
γ	1	0.4		0.3
δ		0.8		0.8
ε	0.5	1	1	
ρ		0.2	0.6	0.7

$=$

	A	B	C	D
a	0.8	0.8	0.5	0.8
b	0.5	1	1	0.7
c	0	1	0.6	0.7
d	0.5	0.5	0.6	0.6
e	0.8	1	0.6	0.7

Let us consider the level $\alpha \geq 0.7$ for the corresponding demand and offer. This results in:

$\underset{\alpha \geq 0.7}{\underline{R} \circ \underline{S}} =$

	A	B	C	D
a	1	1		1
b		1	1	1
c		1		1
d				
e	1	1		1

From here we arrive at the following maximum sub-relations:

$$(b, BCD),\ (ae, ABD),\ (abce, BD).$$

There is no place whatsoever on the team for d. In the figure below we show the corresponding Galois lattice.

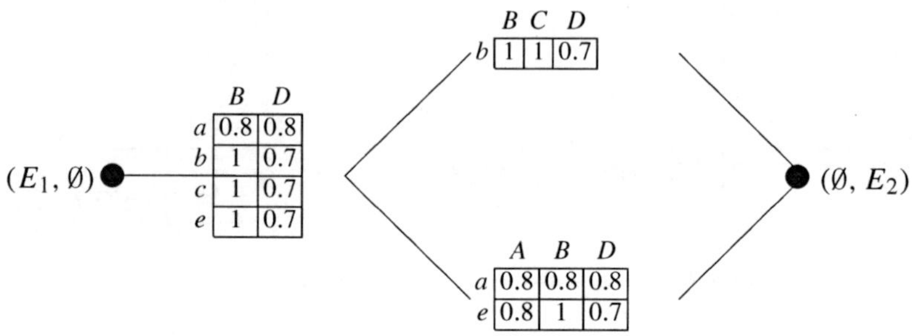

We now do the following assignment:

$$\begin{aligned} a &\longrightarrow D\ (0.8) \\ b &\longrightarrow C\ \ (1) \\ c &\longrightarrow B\ \ (1) \\ e &\longrightarrow A\ (0.8) \end{aligned}$$

This is done by solving the well-known problem of "assignment", which has been around for more than 60 years. To solve this problem, we can use the *Hungarian algorithm* (see, e.g., Gil-Lafuente (2002), pp. 176–193) or dynamic programming. In this case, when all the values are between 0.7 and 1, the assignment with the value 1 can be admitted practically in every case.

The proposed approach could be completed by considering valuations contributed by groups of experts and consideration given to the use of expertons. Expertons form a distributive lattice for $(\vee)$ and $(\wedge)$ but do not form it for the inclusion. In the same way as with confidence intervals, they can be classified by means of their average and, eventually, by another criterion in order to find a partial or total order. Just like with a confidence triplet $(a_1, [a_2, a_3], a_4)$, a total order can be found using the value of $(a_1 + a_2 + a_3 + a_4)/4$ and other criteria. In the case of an experton we propose using its average in order to obtain a total or partial order, so the problem is reduced to what was discussed above.

6 Affinities by Distance in an Arborescence

Next we present a new approach to defining affinities as the basis for the selection of a player. We begin by stating the conditions for a graph to be considered as an arborescence. A directed graph (E, U) is an arborescence with root x_0 if:

1. $\exists x_0 \in E : U^{-1}\{x_0\} = \emptyset$;
2. $\forall x_i \in E, x_i \neq x_0 : |U^{-1}\{x_i\}| = 1$;
3. The graph contains no circuit.

Alternatively speaking, an arborescence is a directed, rooted tree, where all edges point away from the root x_0. Figure 1 provides an illustration.

Since arborescence is a directed acyclic graph, it has a topological ordering, which is an ordering of vertices in a way that the starting endpoint of every edge occurs earlier in the ordering than the ending endpoint of the edge. Thus, it is possible to break down the set of vertices of the graph into disjoint ordered subsets in such a way that if a vertex belongs to a subset numbered k, all the vertices it connects to through an outgoing edge should be placed in a sub-set with a number higher than k. The subsets of such partition are called levels. This can be expressed by means of the following formula. Let us consider a directed acyclic graph (E, U), where we define the sub-sets $N_0, N_1, N_2, \ldots, N_r$ such that

$$\begin{aligned} N_0 &= \{x_i : U^{-1}\{x_i\} = \emptyset\}; \\ N_1 &= \{x_i : U^{-1}\{x_i\} \subseteq N_0\} \setminus N_0; \\ N_2 &= \{x_i : U^{-1}\{x_i\} \subseteq N_1\} \setminus (N_0 \cup N_1); \\ &\vdots \\ N_r &= \{x_i : U^{-1}\{x_i\} \subseteq N_{r-1}\} \setminus \cup_{k=0}^{r-1} N_k, \end{aligned}$$

where r is the smallest integer such that $\cup_{k=0}^{r} N_k = E$. See Gil-Lafuente (2002), pp. 98–111 for more detail. The subsets $N_k, k = 1, 2, \ldots, r$ form a partition of E and are totally and strictly ordered by the relation:

$$N_k \prec N_{k'} \iff k < k'.$$

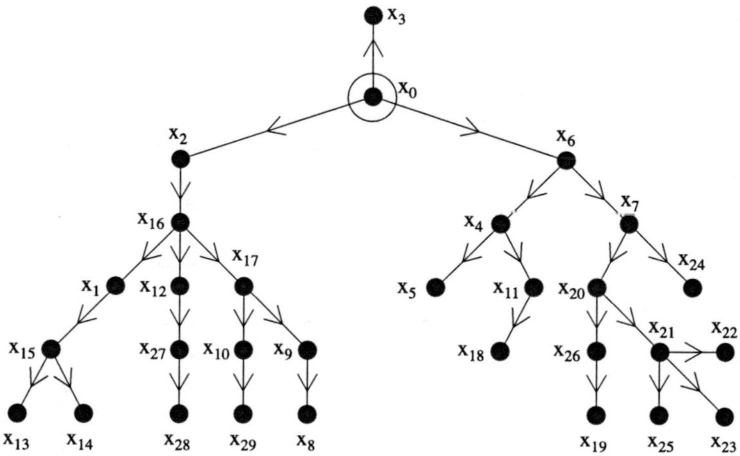

Fig. 1 An arborescence with root x_0

The function $O(x_i)$ defined by

$$x_i \in N_k \iff O(x_i) = k$$

constitutes the ordinal function of the directed acyclic graph. Figure 2 illustrates the concept of ordinal function.

Among many other interesting questions related to arborescence and anti-arborescence, we would like to point out the notion of the distance between vertices. The distance $d(x_i, x_j)$ between two vertices x_i and x_j is the lowest number of arcs in a path between x_i and x_j. Thus, for example, for the arborescence in Fig. 1, we have

$$d(x_{19}, x_{22}) = 4, \;\; d(x_{24}, x_9) = 7.$$

Also of interest is the greatest distance that exists in the arborescence. It can be easily obtained by using the graph of the ordinal function and is equal to two times r, where r is the last level:

$$d_{\max} = 2 \times r$$

In Fig. 2 we have $r = 5$, thus in this case $d_{\max} = 2 \times 5 = 10$.

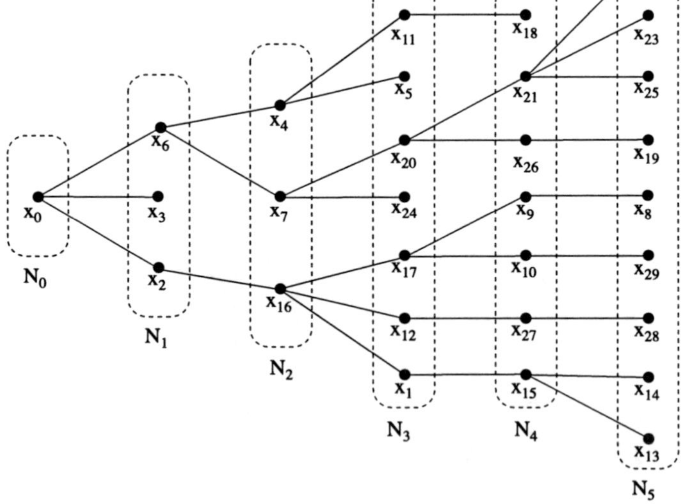

Fig. 2 The ordinal function of the arborescence from Fig. 1

Therefore, a relative distance can be defined:

$$\delta(x_i, x_j) = d(x_i, x_j)/d_{\max}.$$

For example,

$$\delta(x_{22}, x_{24}) = 4/10, \ \delta(x_{21}, x_{13}) = 9/10.$$

Using this idea, we can define an index of affinity or neighborliness:

$$\eta = 1 - \delta.$$

Therefore, for our example we have:

$$\eta(x_{22}, x_{24}) = 1 - 0.4 = 0.6, \ \eta(x_{21}, x_{13}) = 1 - 0.9 = 0.1.$$

The strongest affinity corresponds to 1, that is to say, $\delta = 0$, and the weakest to $\delta = 1$. If $x_i \neq x_j$, we will always have

$$1/(2r) \leq \delta(x_i, x_j) \leq 1.$$

Therefore, we have defined another type of affinity, somewhat different from what we have used in previous sections. This affinity also has a correspondence with several notions defined in topology (neighborliness, etc).

The notion of affinity in arborescence and anti-arborescence is useful in the sporting field for a classification by approximation. It is also possible to generalise this notion by associating with each arc a valuation in [0, 1] that could be either exact numbers or confidence intervals, triplets, etc. Obviously, by using adequate operators so that the notion of distance makes sense, we can define a corresponding affinity index. For the valuation in [0, 1], T-norms or T-co-norms could be used. See Kaufmann and Gil-Aluja (1998) for more detail on these notions.

7 Affinity in a Layered Graph

Many notions in graph theory can be considered, in some way or another, as affinities of different nature (e.g., proximity, classes, subgraphs with specific properties, etc.). The notion of affinity, when taken in a very wide sense, has its correspondence in a diversity of cases in the field of human and social sciences. It is for this reason that we are referring to different concepts that can be inserted into a more general idea linked to the concept of affinity.

The first approach shapes out when attempting to establish the shortest path of a graph. And this is so, because *the further a vertex is from another, the weaker its affinity may be*. The graph represents strong links when the distance in arcs is short, and weak links otherwise.

The alternative definitions of distance between two vertices of a graph are of interest since they provide different definitions of affinity (in the sense of the concepts near or far) passing through a whole gamut of concepts deriving the one from the other. If we only counted the number of arcs, each arc would take the value of 1, and the distance would be the number of arcs covered. In this case, the number of arcs between two vertices x_i and x_j is divided between the ordinal function less 1 in order to arrive at an index that belongs to [0, 1].

Let us conclude by saying that that this is an aspect of special importance very worthy of our attention for its development in the future, as we study the complexity of the selection of players for a team game.

References

Aristotle, (1977). *Obras. Lógica. De la expresión o interpretación*. Barcelona: Aguilar Publishing.

Bergson, H. (1970). *Le possible et le réel*, en Oeuvres. Paris: PUF, Edition du Centenaire.

Clausius, R. (1865). *Annals of Physics*, CXXV.

De Luca, A., & Termini, S. (1972). A definition of non-probabilistic entropy in the setting of fuzzy sets theory. *Information and Control, 20*, 301–312.

Gil Lafuente, J. (2002). *Algortimos para la excelencia*. Claves para el éxito de la gestión deportiva. Vigo: Publ. Milladoiro.

Gil-Aluja, J. (1998). *Elements for a theory of decisions in uncertainty*. Dordretch: Kluwer.

Kaufmann, A., & Gil Aluja, J. (1993). *Técnicas especiales para la gestión de expertos*. Santiago de Compostela: Publ. Milladorio.

Lukaciewicz, J. (1910). O zasadzie wylaczonego srodka. *Przegl'd Filozficzny, 13*, 372–373.

Moles, A. A. (1990). *Les Sciences de l'imprécis*. Paris: éditions du Seuil.

Poincaré, H. (1906). *La science et l'hypothése*. Paris: Flammarion.

Prigogine, I. (1997). *La fin des certitudes*. Spanish translation titled El fin de las certidumbres. Buenos Aires: Publ. Taurus.

Rosenfeld, L. (1971). Fuzzy groups, *Journal of Mathematical Analysis and Applications, 35*, 512–517.

Sugeno, M. (1977). Fuzzy measures and fuzzy integrals, a survey. In M. M. Gupta, G. N. Saridis, & B. R. Gaines (Eds.), *Fuzzy automata and decision processes*. Amsterdam: North-Holland.

Zadeh, L. (1965). Fuzzy sets. *Information and Control, 8*, 338–353.

Zadeh, L. (1996). Fuzzy logic and the calculi of fuzzy rules and fuzzy graphs. *International Journal of Multiple-Valued Logic, 1*, 1–39.

Zimmermann, H. J. (1978). Results of empirical studies in fuzzy set theory. In G. J. Klir (Ed.), *Applied general systems rechearch*. New York: Plenum Press.

Game Systems in Team Sports

Jaime Gil-Lafuente

Abstract Spectator team sports are now being studied more frequently by those who have, or wish to have, responsibility of the smooth functioning of club sports. Technical directors, managers, trainers, and individuals in other similar roles look for "formulas" that can help them to obtain good qualifying results and present in-person spectators and the television and general audience with an interesting and visually stimulating show. Recently, the "game system" concept has become popular, providing a small set of formulas of attack and containment capable of giving unique quality or a distinct identity to a team. Because terms like "game system" are often repeated and circulated by word of mouth without precise knowledge of what they really mean, we attempt to establish a definition for this term. This definition, in turn, will provide the basis for a methodology for neutralizing the problems that impede or hinder the achievement of the objectives sought through the development of the game.

1 Concept and Content of Game Systems

We deem the "game system" a homogenous set of circulatory diagrams of the ball (or equivalent object) that can be represented by networks or arborescence (as well as anti-arborescence) whose objective is to ensure the best possible performance of the available athletes through victories over their opponents. The proposed definition includes three fundamental elements:

1. A homogenous set of circulatory diagrams. The notion of the "system" in sports has to do with characteristics that differentiate one team's method from other systems or simply from other ways of playing the game. Thus, it seems logical to think about the existence of a basic "method" to make the ball move (for example), from which there can exist variants that are adopted or not according to events on a particular game or during a specific competition.

J. Gil-Lafuente (✉)
Department of Business Administration, University of Barcelona, 08034 Barcelona, Spain
e-mail: j.gil@ub.edu

S. Butenko et al. (eds.), *Optimal Strategies in Sports Economics and Management*,
DOI 10.1007/978-3-642-13205-6_2, © Springer-Verlag Berlin Heidelberg 2010

2. Representability through networks or arborescence. When one is attempting an in-depth study of a team's plays, circulatory diagrams as a form of visualization are almost essential. Thus, as we frequently see in sports television programs, a trainer will show on a blackboard or screen how the ball should circulate and/or how the players should move. Reticular diagrams and arborescence (and also anti-arborescence) are among the tools currently used for this purpose.
3. Success in sports. The quest for overall success through victory in each contest, game or match is integral to the spectacle of sports. Normally, victory is achieved when a team attains the greater number of points (e.g., through goals or baskets), while blocking or limiting those of its opponent. Circulatory diagrams should provide "lines" of movement by which offensive players can achieve goals and those by which they can intercept the opponent.

These reflections on the definition of the "game system" have permitted the introduction of other concepts that will be the object of subsequent investigation. We refer to networks, sub-networks, schemes, and cuts, all of which we attempt to introduce into a unitary model.

2 Game Diagrams

Having clarified this fundamental terminology, we will move on to a much more practical phase, situating ourselves in the position of the trainer or an analogous person—who, before the athletes "executed" their instructions, plotted on the board a positional diagram of some (or all) of the players and indicated how the ball should circulate.

First, with the object of beginning the study in the simplest manner possible, we will suppose that the sports official, manager or trainer (we will use this term for simplicity) establish that the ball should only be "touched" once by each athlete and that a maximum of just three athletes should be involved in a play. This is not an unreasonable scenario to begin with, and it allows the introduction of more complex diagrams.

As is well known, the possible diagrams form a free distributive lattice with three generators. The number of elements in this lattice is 18. Hence, that is also the number of possible diagrams. Before representing the lattice and the 18 diagrams from which it is derived, we must remember that the number of elements of the distributive n-lattice with generators increases rapidly as n increases. This soon makes it practically impossible to represent all possible diagrams. Thus, for example, when $n = 2$, the number of possible diagrams is four; if $n = 3$, then the number of possible diagrams is 18, as we have noted; when $n = 4$, now the number of diagrams is 166; when $n = 5$, the number of diagrams rises to 7,578; and finally, when $n = 6$, the number of possible diagrams becomes extremely high at 7,828,352. Obviously, this study will not consider every possibility, but will instead choose from among them according to their utility, highlighting

Fig. 1

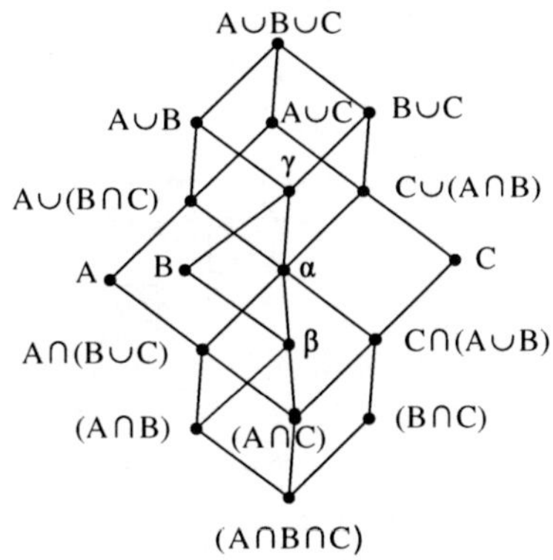

$$\alpha = (A \cap B) \cup (B \cap C) \cup (C \cap A)$$
$$\beta = B \cap (A \cup C)$$
$$\gamma = B \cup (A \cap C)$$

those that are likely to be adopted in reality and by their homogeneity in leading to a "game system". Granted these observations, we will continue with our simple example, presenting the network for three athletes (or three generators), see Fig. 1.

Established clarifications in relation with the possibility to examine all or part of the diagrams to be developed by my team, we will go through the representation of each one of the more useful and pertinent ones. This might seem to call for establishing the theoretical basis on which the concept that we propose is based, but for educational purposes, we will first continue presenting our case, leaving the theoretical justifications for later.

The 18 vertices of the grid show the 18 possible diagrams with three players who only have the ball once in the game. We will assign to these three athletes the capital letters A, B, and C. When the operator $\cup$ appears in the grid, the players are acting in "parallel" (one player or the other plays) and when $\cap$ appears, the players are acting in a "series" (one player plays "and" then the other does). We can now represent the 18 diagrams using the reticular/grid formula. With the objective of maintaining consistency with the grid, we are going to arrange the diagrams in the same position as the vertices of the network.

At each end of the network, one sees the beginning of a play (in soccer, it could be the ball being played out by the goalkeeper) and the end of the play (for example, the ball reaching the other goal). These parts of the play are designated by I and F, respectively (see Fig. 2).

Based on these networks, it is possible to formulate certain pertinent questions.

1. Not all 18 plays merit the attention of the sports manager. Some will perhaps consider them unviable, while others, although they think them possible, will find that they are not consistent with what is perceived of as "modern play".

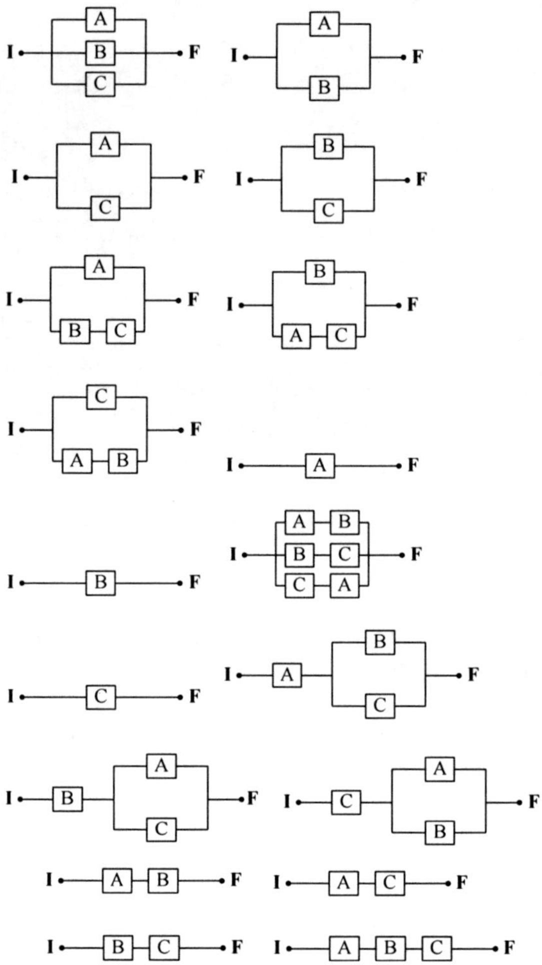

Fig. 2

Fortunately, because of this, there are distinct "game systems". Thus, out of all possibilities, the sports official will chose the systems that best fit his way of thinking.

2. The placement of two players "in order" (in a horizontal line)—for example, with *A* first and *B* afterwards—does not predetermine their order. The same diagram is valid with *B* first and *A* afterwards in our example. This principle of interchangeability is valid for any players who act in series, but not when they act in parallel, acting one "or" the other.
3. The placement of two players in the same line does not require them to run or move in a straight line; rather, the ball passes from one to another. We have drawn the networks using rectangular figures to remain consistent

with the reticular analysis done in other fields, but this does not mean that the athlete has to follow a pattern of precisely angulated lines. It would be perfectly acceptable, and in many cases desirable, for example, to follow a straight line towards F. What is important is maintenance of the sequence of players.

4. It will be observed that there does not appear to be any sign that indicates the direction of the game in any of these networks. Thus, the ball can circulate either from I to F or the reverse (forwards and backwards). We will tell that in theory, oriented arches/arcs do not exist under these circumstances. However, it is always possible to introduce this element. Then, the associated lines will become arrows, and the players will only be able to circulate in the direction indicated by the arrows (also called arcs/arches). In a network, there, for example, may be one oriented arc/arch among others without any specific orientation (that allow circulation in one direction and the other).

To move from the descriptive to the operational realm, we introduce a more complex diagram (reticular representation) than those shown so far, including five athletes: A, B, C, D, and E. We initially maintained the hypothesis that the ball only passes once to each of them. The manager, after adequate advice, presents the network shown in Fig. 3.

With this approach, the objective of the sports official is to determine the ball's shortest "path" from I to F or to avoid its going from F to I. For the former, this means examining all of the possibilities that the proposed diagram offers, the "minimum ties" that permit the advancement of the ball from the origin of the play to the end.

In this simple example, it appears that there are at least three possible schemes: $\{C\}, \{D, E\}$, and $\{A, B, E\}$. Graphically, we can represent these as in Fig. 4.

On the other hand, if it were the rival team's trainer who were using this diagram, the sports official for our team would be working to block those planned movements. In other words, the objective then would be to "cut" the planned movements. The task is now to find the "minimum lengths" that achieve this.

In this simple scenario, one can find the minimum lengths (which are as follows) by sight:

$$\{C, D, A\}, \{C, D, B\}, \text{ and } \{C, E\}.$$

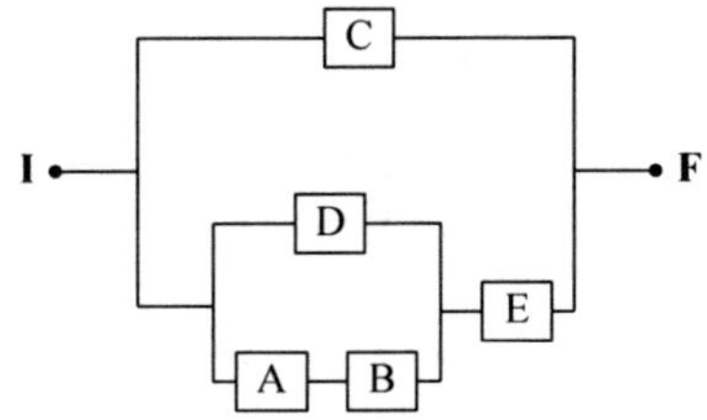

Fig. 3

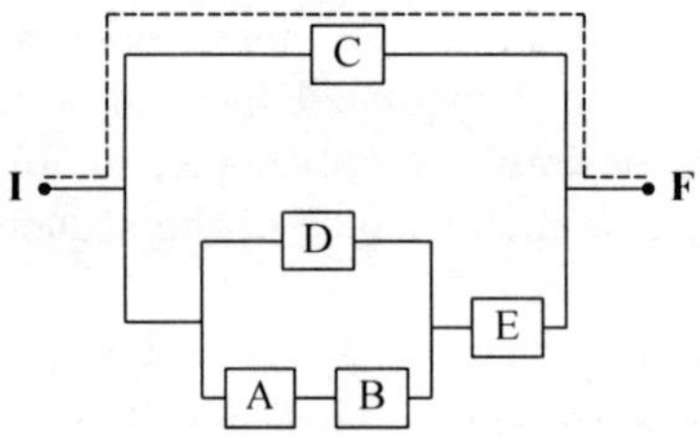

The first minimum scheme

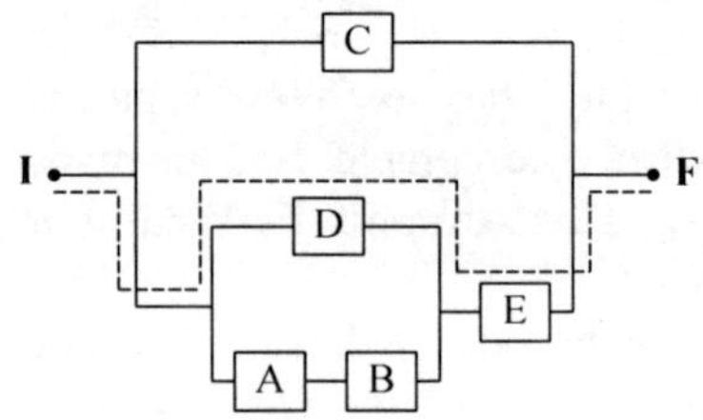

The second minimum scheme

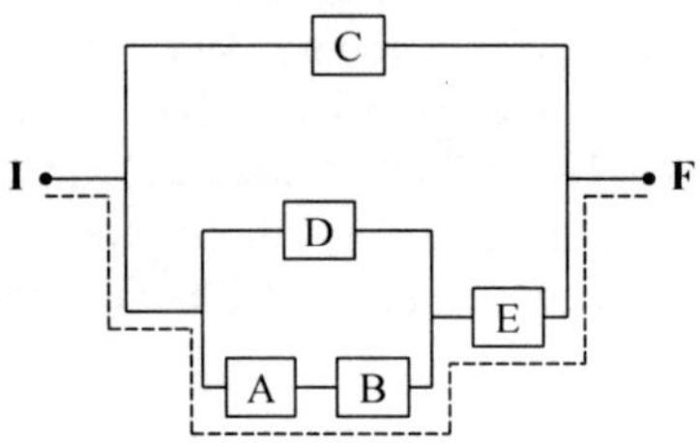

The third minimum scheme

Fig. 4

Graphically, we can represent these as in Fig. 5.

In this first example, intended as an introduction to a possible theory of game systems, two fundamental concepts have appeared: those of "minimum ties" and "minimum lengths".

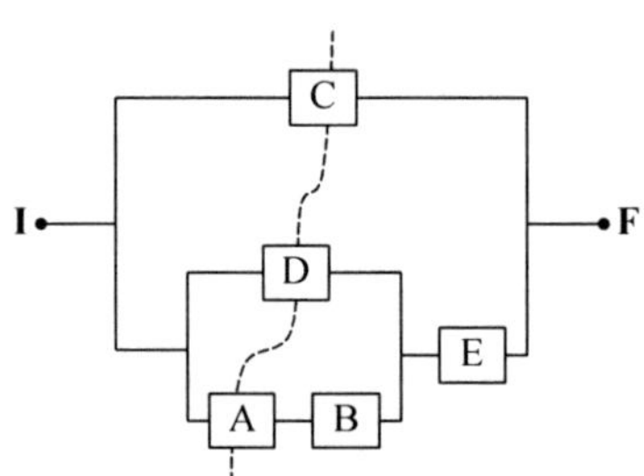

The first minimum length

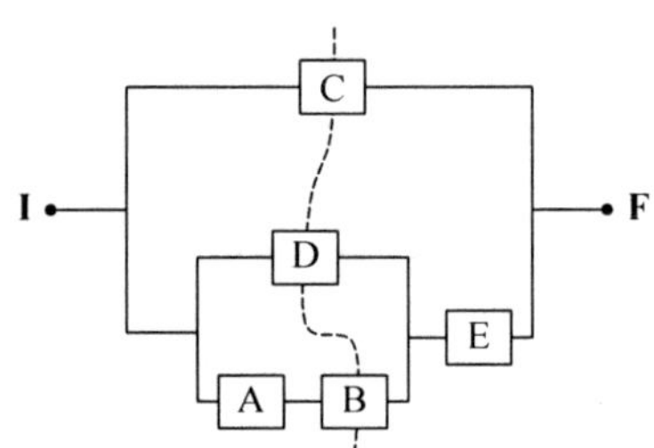

The second minimum length

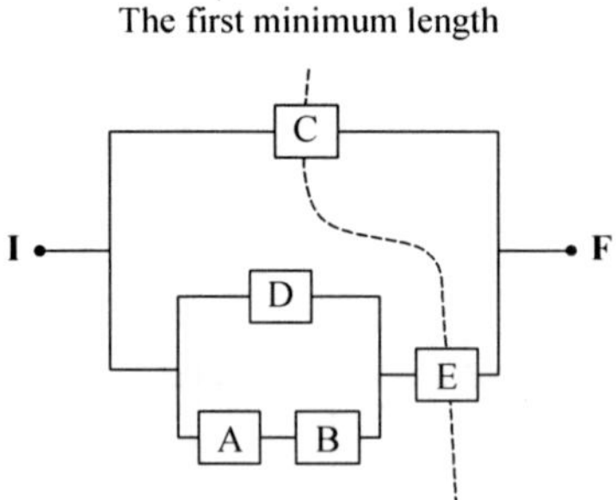

Fig. 5

Fig. 6

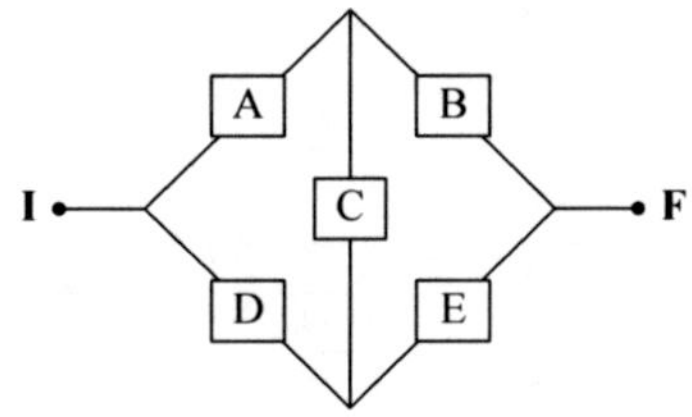

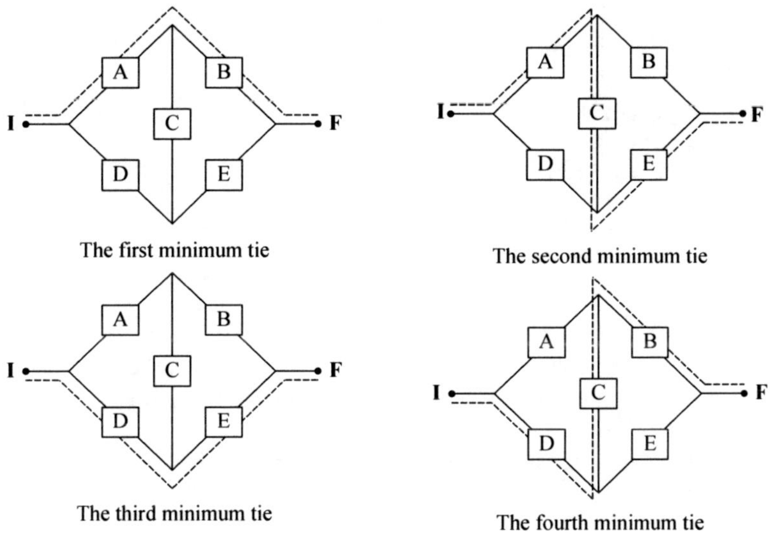

Fig. 7

Considering the same basic construction, we present another example. Here, the same players assume different positions, such that the circulation of the ball will also be different. The suggested diagram is as in Fig. 6.

The first objective is to develop an offensive play. There are now four "minimum ties":

$$\{A, B\}, \{A, C, E\}, \{D, E\}, \text{ and } \{B, C, D\}.$$

We can graphically represent these as in Fig. 7.

For the second objective, defensive action, there are also four "minimum lengths":

$$\{A, D\}, \{B, E\}, \{A, C, E\}, \{B, C, D\}.$$

See Fig. 8 for an illustration. It is thus observed that if the team playing defense can successfully block the marked players through the discontinuous line in each network, the opposing team's play is destined to fail. We can now ask ourselves

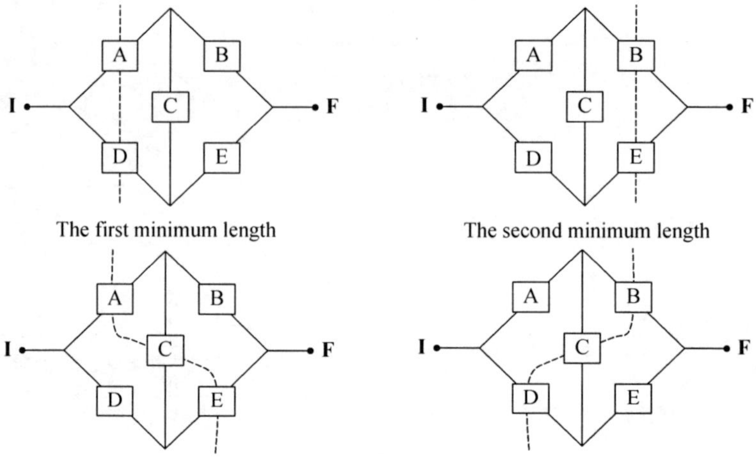

Fig. 8

what happens when, because of a player's limitations or habits, the movement of the ball is limited to one direction. To consider this possibility, we can use the same diagram but assume that player C can only move the ball in the direction marked by the arrow noted in Fig. 9.

This condition significantly changes the situation. In effect, in terms of offense, the four available minimum schemes have been reduced to the following three:

$$\{A, B\}, \{A, C, E\}, \text{ and } \{D, E\}.$$

In our illustration, there would no longer be a fourth minimum tie. Something similar happens with defense. The four minimum lengths have been reduced to the following three:

$$\{A, D\}, \{A, C, E\}, \text{ and } \{B, E\}.$$

In our illustration, the fourth minimum length would disappear.

With these examples, we believe to have demonstrated the practical meaning of two fundamental concepts: those of "minimum ties" and "minimum lengths". However, we should also note an underlying hypothesis: "If a player can make a

Fig. 9

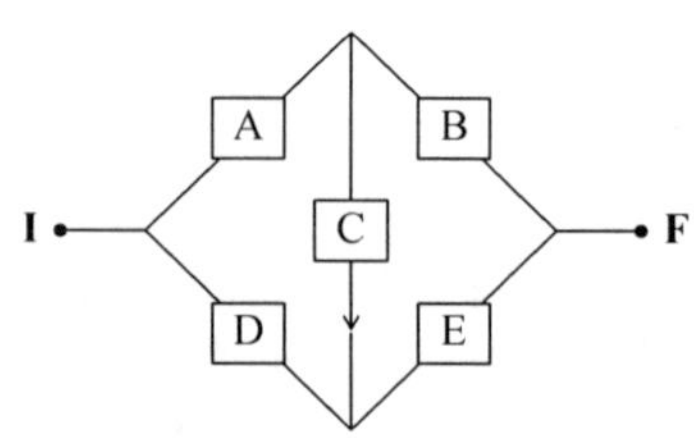

play himself, he will not utilize other players". This hypothesis is based on the idea that in the game, taking place with the same number of players on each team, if the player who has the ball is free (of marking), his teammates will not be free; therefore, it assumes an unnecessary risk by involving another marked player.

When a player is marked and unable to escape, then the intervention of another player appears to be necessary. Despite this, it is also possible to dispense with this assumption (which we have now made explicit), although we are going to maintain it for our purposes, so as not to break the thread of the exposition.

3 Generalization of the Concepts of Our Diagrams

Until now, we have focused on diagrams in which each player intervenes only once in each play. Now we will consider a model that allows players, if applicable, to occupy another position on the court or field once the ball is released such that they can receive it again. This represents a type of generalization of previously presented player involvement. Again, we will begin with an example shown in Fig. 10.

In this network, it can be observed that players A, B and C now move from their original "lines" to other positions; they have the potential to touch the ball on more than one occasion and with or without the intervention of another player. In this diagram, the offensive position allows the minimum schemes $\{A\}$, $\{B\}$, and $\{C, D\}$, where we can visualize these via the graphics as in Fig. 11. For the defensive position, two minimum lengths appear, $\{A, B, C\}$, $\{A, B, D\}$, whose representations in the network are as in Fig. 12. Observe that, in the first diagram, we have shown the "cut line" in the more advanced positions of $\{A, B, C\}$. We would also have been able to do this in the rearmost positions without weakening the effectiveness of the exposition. As such, this can be considered: three cuts instead of the two that we have noted. Based on this new assumption, the cut $\{A, B, C\}$ would repeat in two configurations: one more advanced in the diagram than the other. In reality, there are three players who move forward and backwards in an area of the field or court.

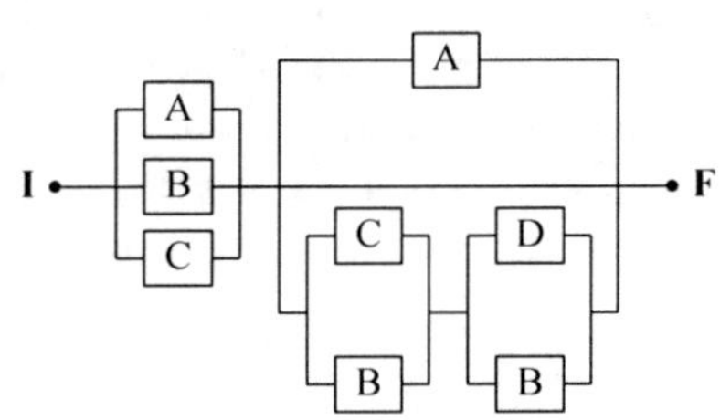

Fig. 10

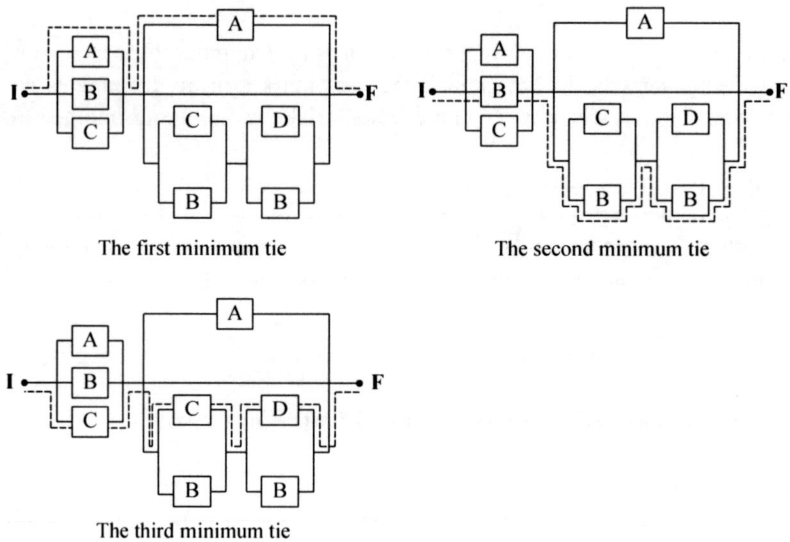

The first minimum tie

The second minimum tie

The third minimum tie

Fig. 11

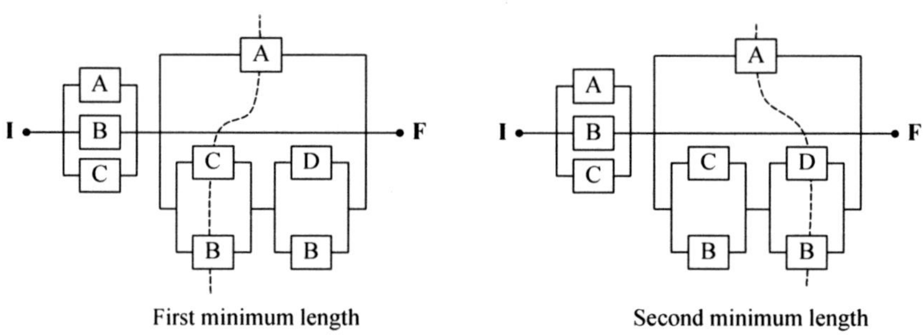

First minimum length

Second minimum length

Fig. 12

4 Paths and Cuts in Complex Diagrams

In the previous sections, we have shown, given certain assumptions, some possible diagrams of plays, all of them including 3, 4 or (at most) 5 players. However, there are few team sports in which the teams are this small. At the beginning of this paper, we highlighted that increasing the number of players increases the number of possible different diagrams astronomically. Thus, in assumptions that limit this

Fig. 13

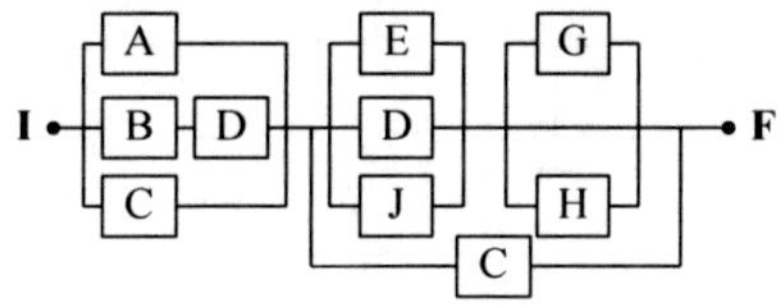

number (for example the non-repetition of a player in more than one position), the problem of how to operationalize a system with an increased number of players arises. For example, in soccer or football, the team will consist of a goalkeeper and ten more players. We have found a solution for use in the context of this and other sports with a similar number of active participants: American football, basketball and rugby, among others. The key is to decompose and compose global diagrams represented by networks. In more technical terms, the goal is to decompose a network into sub-networks and then reconstitute it. Fortunately, we know the proper techniques for doing this, sometimes with only minimal study of the diagrams. Below, we will present several examples. We begin with a complex network as in Fig. 13. This network includes eight players, a number sufficient to study this process of decomposition. In this diagram, one can observe that there are nine minimum offensive ties: $\{C\}$, $\{A, D, G\}$, $\{A, D, H\}$, $\{A, E, G\}$, $\{A, E, H\}$, $\{A, J, G\}$, $\{A, J, H\}$, $\{B, D, G\}$, and $\{B, D, H\}$. We graphically present these in Fig. 14.

The description and graphic representation of these minimum schemes encourages certain reflections. First of all, the number and complexity of the minimum schemes have increased even though we have limited ourselves to 8 players and only 2 of them engage in "repeat play": players C and D. Secondly, it is difficult to visualize all of the minimum ties without omitting some or making mistakes. We approach the first issue by decomposing the network into subnetworks. Regarding the second, we can simply say that there should be no problem in analytically determining the minimum schemes. We will address this question in greater depth later on but for the moment move on to consider defensive action.

Here, there are four minimum lengths at play as follows: $\{A, B, C\}$, $\{A, C, D\}$, $\{C, D, E, J\}$, and $\{C, G, H\}$. They can be graphically represented as in Fig. 15. It can be observed that the first and second minimum lengths differ only in one player. In the first play, the "cancelation" must be at B, whereas it must be at D in the second play. One should note, however, that the two players are in line at the same place in the diagram; this seems to indicate that, for defensive purposes, there is a "redundancy" and therefore that one of them is superfluous. Nevertheless, leaving aside this issue for the moment, we see what can be done to simplify the work of analyzing the diagrams as they are represented here.

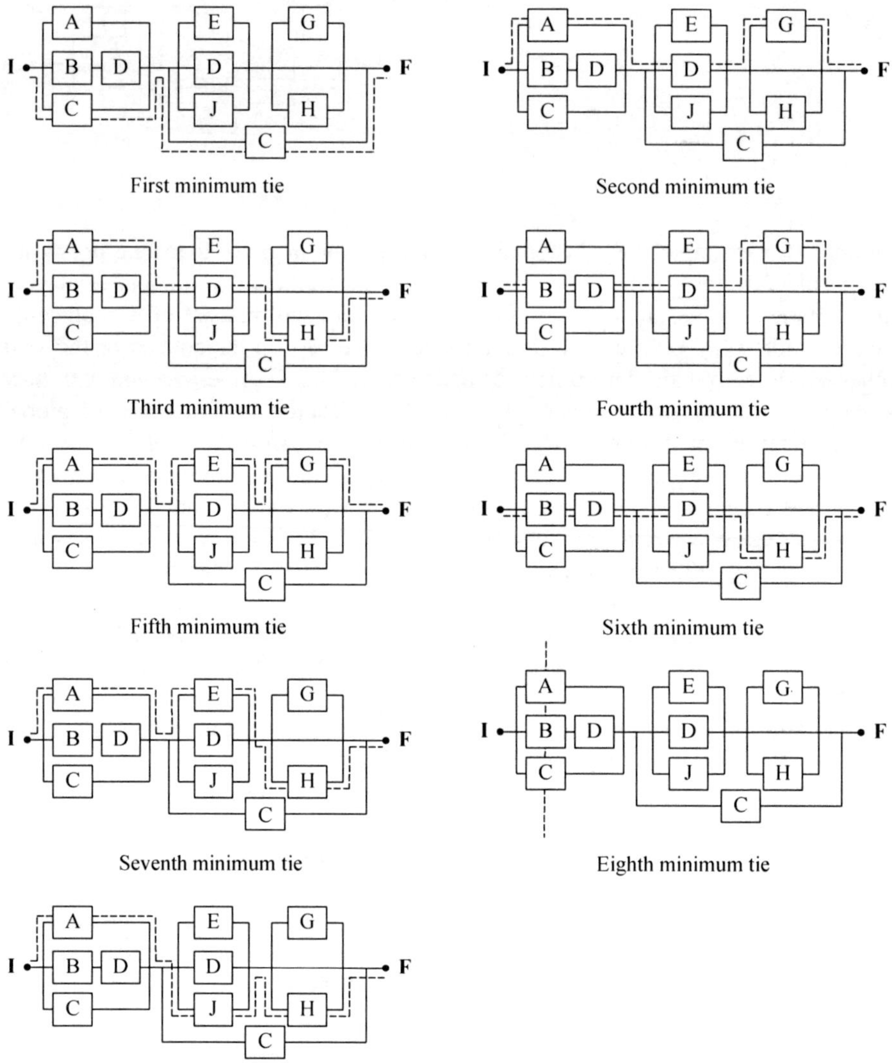

First minimum tie

Second minimum tie

Third minimum tie

Fourth minimum tie

Fifth minimum tie

Sixth minimum tie

Seventh minimum tie

Eighth minimum tie

Ninth minimum tie

Fig. 14

5 Partial Analysis of a Global Diagram

Every complex diagram represented by a network/chain can be decomposed into various partial diagrams, each one of which can be conceptualized as a sub-network. In addition, representing a pair of theoretical rules, specified exactly with two operators, if they later want to return to form the initial network. Of course, there are various ways of decomposing the network; the same diagram of a play can yield various groups of sub-diagrams. In other words, according to necessities or desire,

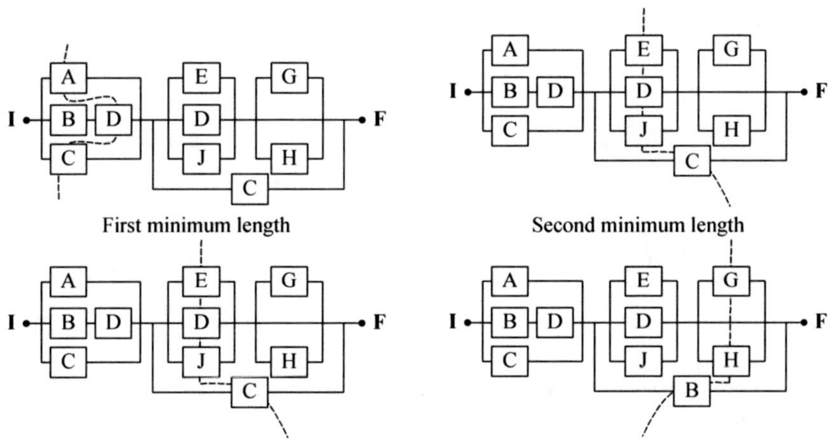

Fig. 15

the network can be cut into pieces in a distinct manner, and all of the possibilities are theoretically correct. In practice, the sports official will choose the method that seems more useful for the team's particular purposes.

Given this word of caution, we suppose that the trainer wishes to separately analyze the 4 sub-networks denoted by S_1, S_2, S_3, S_4 that appear in Fig. 16. As we have just noted, this can be considered a network of sub-networks. In effect, after theoretical merging of the components of a subgroup, the players of a sub-network act as just one player, as shown in Fig. 17. This is very useful—above all, for the recomposition of a network from sub-networks as we will see in the following. Based on this decomposition, it is possible to separately study each of the four sub-networks. We will begin with sub-network s_1, which is graphically represented as in Fig. 18. It is easy to establish the minimum schemes (offensive position):

$$\{A\}, \{B, D\}, \{C\}.$$

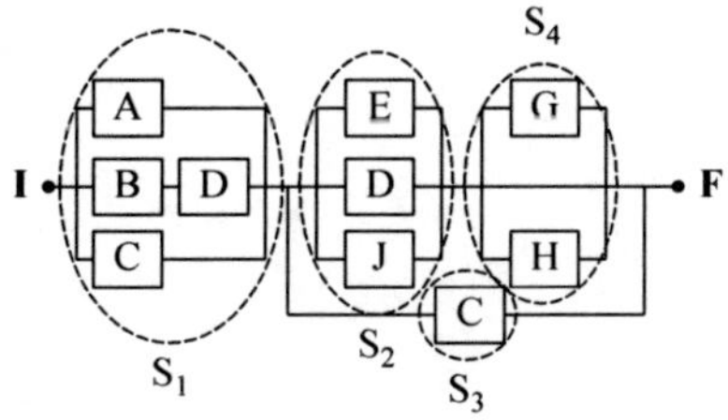

Fig. 16

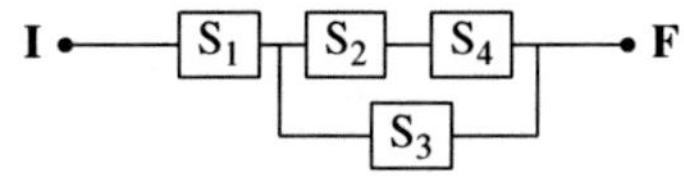

Fig. 17

Fig. 18

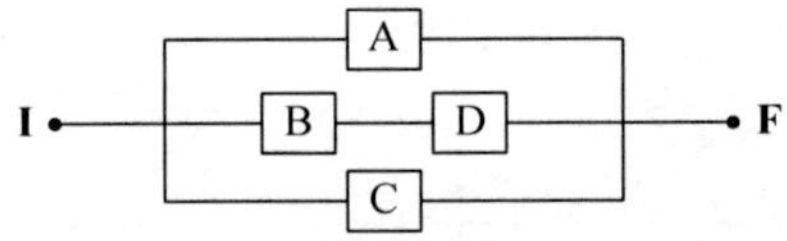

The same is true of the minimum lengths:

$$\{A, B, C\}, \{A, C, D\}.$$

Thus, it becomes very easy to make a partial study of this zone of play. It will not escape the reader that these segmented analyzes are helpful in devising positional changes or modifications to the structure of the sub-network. The study of the other sub-networks, which is presented below, is even easier.

Sub-network S_2 is shown in Fig. 19. There are three minimum ties—$\{E\}, \{D\}$, and $\{J\}$—with only one minimum length: $\{E, D, J\}$. Sub-network S_3 is shown in Fig. 20. Here, there is evidently a minimum tie and a minimum length that coincide: $\{C\}$. Sub-network S_4 is shown in Fig. 21. There are two minimum ties, $\{G\}$ and $\{H\}$, and there is only one minimum length, $\{G, H\}$.

We have now analyzed the component parts of the global network. However, one cannot automatically fuse the partial minimum schemes and cuts into a global system of minimum schemes and cuts. In this phase of the study, we will determine them taking into account the entire network.

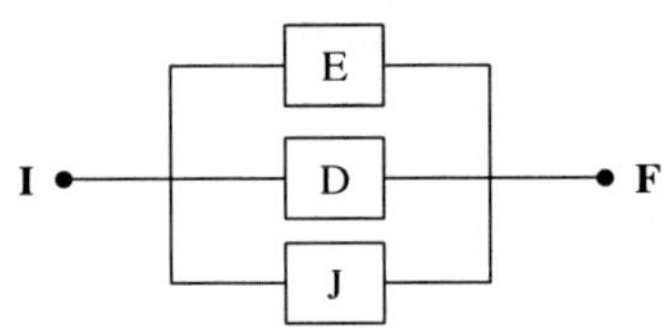

Fig. 19

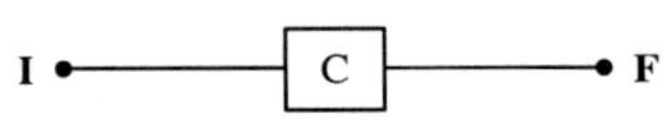

Fig. 20

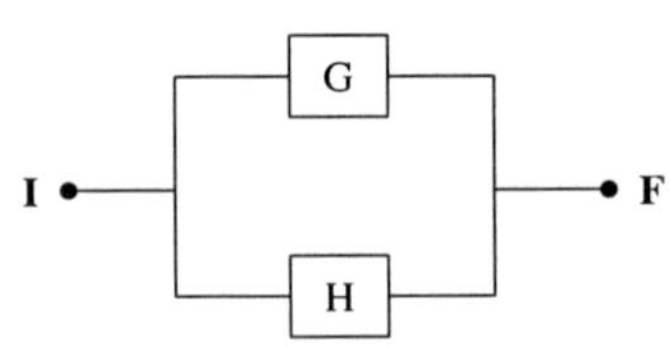

Fig. 21

6 Modification of the Diagrams for the Game

Suppose that the sports official wishes to modify the diagram through the sub-network S_1-displacing, for example, players B and D to move them to a new diagram, as is represented in Fig. 22. Once available, the new S_1 sub-network recombines with the others to help create the global network as previously stated. We have the diagram shown in Fig. 23. However, to modify the structure of one or more sub-networks is not the only useful option. Additionally, the sports official might decide to rearrange the network or sub-networks themselves. Thus, for example, the prior network of sub-networks shown in Fig. 24 could become the network shown in Fig. 25. Then, the general network looks as in Fig. 26.

For this network to be valid, one must reevaluate the minimum ties and cuts, which normally do not coincide with those in the existing network before the adjustments to the sub-networks. A question can be posed in relation to these changes. The change between S_1 and S_2 (with independence of the variation in the positioning of S_3) means, in a certain sense, moving players from mid-field or mid-court (those that form S_2) to defensive positions and, conversely, moving defensive players to positions at mid-field or mid-court, and this is not always possible in all team sports. The question to ask is what would happen if the players in the sub-networks interchange/exchange and the only thing that varied was the relative positions of the sub-networks? Evidently, not much would change. The same general network would

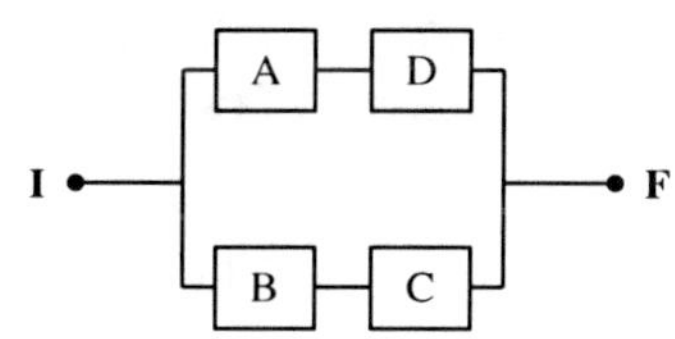

Fig. 22

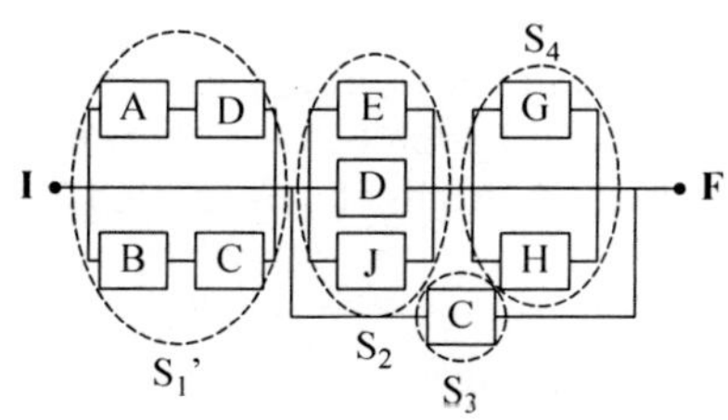

Fig. 23

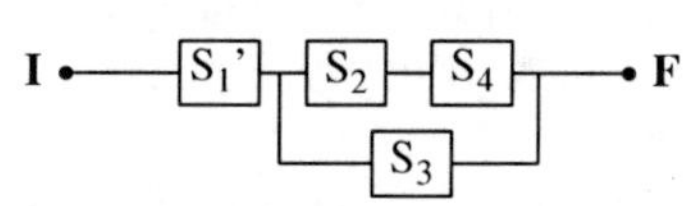

Fig. 24

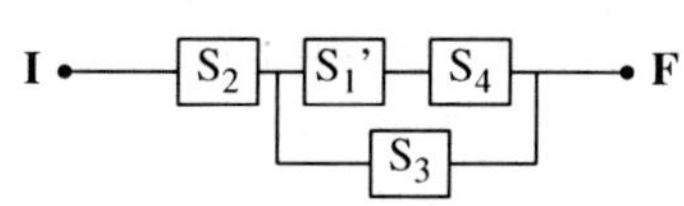

Fig. 25

Fig. 26

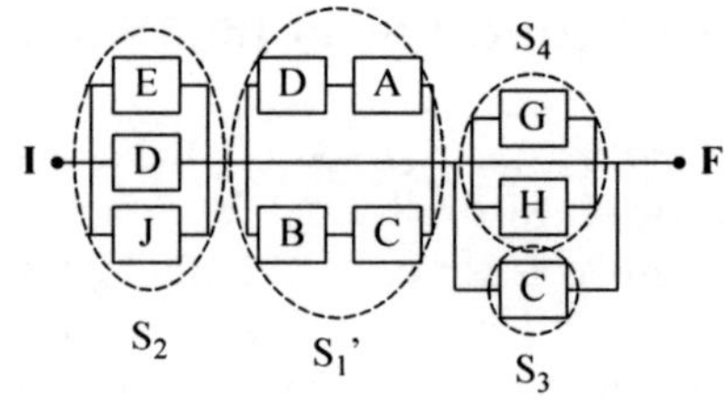

exist, and the only modifications would be changes to which players were in what position. Easier still, one might preserve the original configuration of letters but just assign them to different players, as appropriate.

7 Study of the Game by Zones

To end our examination of this subject, we reverse our initial approach and assume that the trainer wishes to study the game based on the zones of the field or court, creating partial diagrams that can later be conveniently integrated into a general illustration. This process is similar to the last one we studied, but with small variations (the construction of the larger network occurs after the formation of each sub-network). We will consider this problem using an example. Let us say that the problem consists of two perfectly differentiable phases:

(1) Construction of the sub-networks
(2) Implementation of a network

Let us suppose that a trainer wishes to conduct a study of the circulation of the ball, dividing the field or court into four zones: Z_1 (rear or defensive), Z_2 (left of center), Z_3 (right of center), and Z_4 (front).

There are 11 players: 4 in zone Z_1, 2 in zone Z_2, 2 in zone Z_3, and 3 in zone Z_4. The sports official, after the corresponding analysis, decides to establish the diagrams shown in Fig. 27, represented by the respective sub-networks.

In constructing these sub-networks, it is not our intention to provide a real diagram that would routinely be used within a game system. We simply wish to present a totally arbitrary example. We now move to the second phase, constructing the network of sub-networks (see Fig. 28). In this, as in many other cases, the formation of the general network is immediate, see Fig. 29.

It is this network that one references when one needs to ascertain the minimum ties for the offensive positions and the minimum lengths for the defensive positions. This can be done independently of the partial analyzes that can be studied separately for each one of the 4 sub-networks.

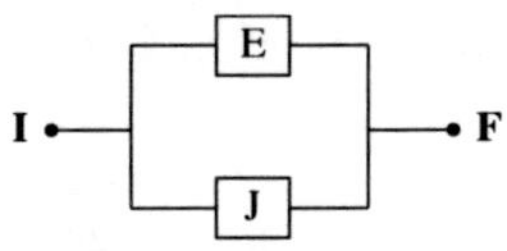

Sub-network S_2 of Zone Z_2

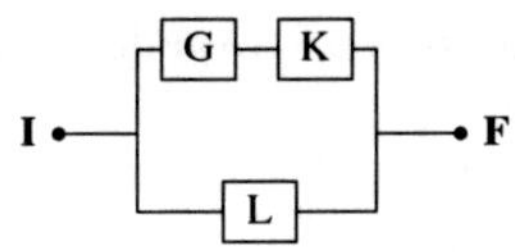

Sub-network S_4 of Zone Z_4

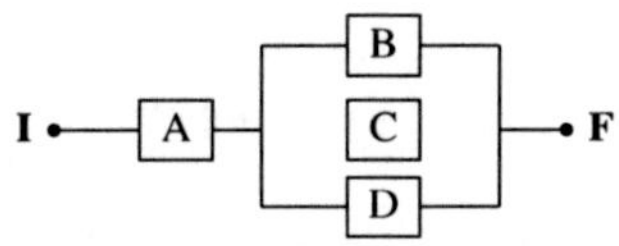

Sub-network S_1 of Zone Z_1

Sub-network S_3 of Zone Z_3

Fig. 27

Fig. 28

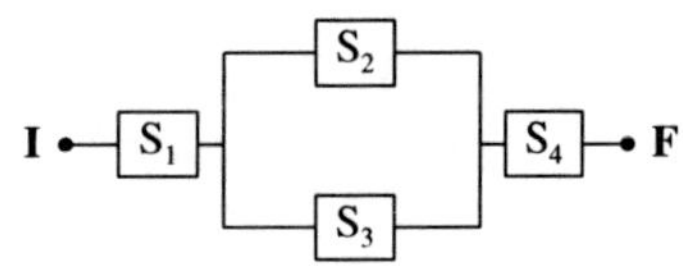

Fig. 29

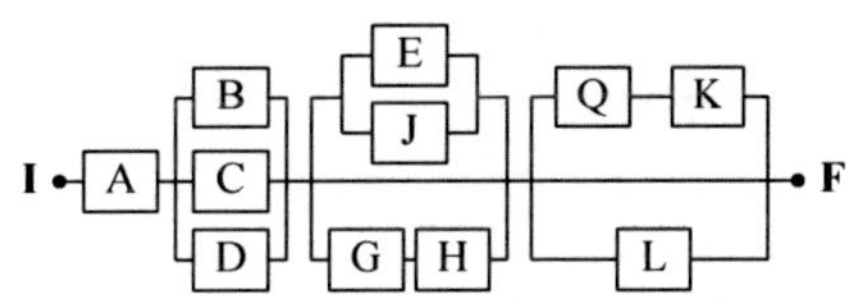

8 The Function of the Structure of a Game Diagram

We have presented numerous examples of the construction and functioning of diagrams and sub-diagrams of plays, avoiding mathematical formulations until now. The decomposition and recomposition of a network into sub-networks and the various sub-networks in a general network have been our object of analysis based on simple logic. We have determined the minimum “ties”, “lengths”, or “cuts” in a simple visual manner, trusting in the intuition and high-quality work of sports officials. However, we believe that while this approach is interesting, a scientifically rigorous analysis of the subject requires us to create technical methods of automatically finding such solutions without the possibility of error or omission.

Furthermore, some urgent questions arise. Among them is the reliability or unreliability of the game system or of the diagrams of plays. One can and should reflect on what happens if one of the players fails to execute a particular move during an encounter. In addition, one must of course attempt to optimize the functioning of the game system. Furthermore, other lesser questions deserve special attention. For example, when a player is expelled from the field or is hurt without a substitute being called in, the diagram of the play must be recomposed and the positions adjusted in the best way possible. It is important to know if the presence of the player on the field is superfluous because there are one or more players who are already doing the work that this player was assigned to do (how many times we have seen or heard that two players are "in the way"?). At times, the value of a player on the rival team makes particular attention necessary. In this case two players could mark the same player. These decisions, among others, have the potential to assure the success or failure of a team.

It is for these reasons that we will undertake this part of our study, considering the offensive position of the player, whom we denote as A. This individual may or may not perform adequately (they may have significant weaknesses). Therefore, we assign a binary variable to the functioning of A such that

If A yields, it will equal 1
If A does not yield, it will equal 0

We next consider the simplest case, formed by 2 players: A and B. Each one is assigned a binary variable, a for A and b for B. Only two diagrams can be formed that include only these two athletes.

1. If A yields "and" B yields, system S functions. The network, already known as the "network in series", is shown in Fig. 30.
   If one of the two does not yield the diagram no longer functions. Let us imagine all of the possible cases:
   A yields "and" B yields; S functions;
   A does not yield "and" B yields; S does not function;
   A yields "and" B does not yield; S does not function;
   A does not yield "and" B does not yield; S does not function.
2. If A yields "and/or" B yields, the system S functions. The network, also known as a "parallel network", is shown in Fig. 31.

Fig. 30

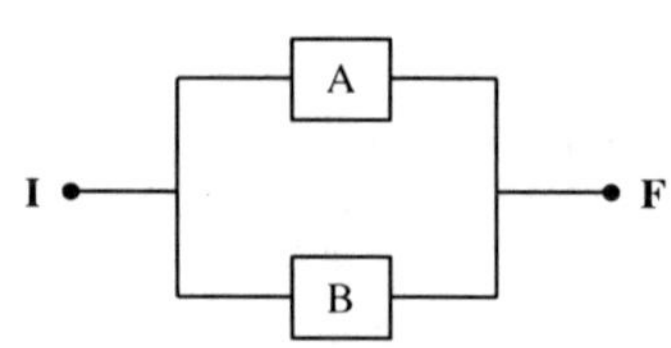

Fig. 31

It is sufficient then that one of the two players yields (although both may also do so) for the system to function. The possible cases are as follows:

A yields "and" *B* yields; *S* functions;
A does not yields "and" *B* yields; *S* functions;
A yields "and" *B* does not yield; *S* functions;
A does not yield "and" *B* does not yield; *S* does not function.

It can easily be proved that the only operators capable of faithfully reflecting the two diagrams are $(\cdot)$ and $(+)$. In effect, if a function is formed that represents these two situations and we call it $\varphi(a, b)$, then for the network in series

$$\varphi(a, b) = a \cdot b,$$

where the symbol $\cdot$ (product) semantically reflects the "and", and for the parallel network

$$\varphi(a, b) = a + b,$$

where the symbol $+$ (algebraic sum) semantically corresponds to "and/or".

Thus, it can be said that every diagram of play represented by a network possesses its own function φ that is called its "structural function". Thus, all of the previously presented networks have their own structural function.

In general, when various athletes $A, B, C, \dots, K$ are involved in a play, the function of the structure will be depicted as follows:

$$\varphi(a, b, c, \dots, k).$$

Below is a key example of what form the structural function of some of the networks previously studied takes. We begin with the diagram shown in Fig. 32. The function of the structure is as follows:

$$\varphi(a, b, c, d, e) = c + (d + a \cdot b) \cdot e$$
$$\varphi(a, b, c, d, e) = c + de + a \cdot b \cdot e$$

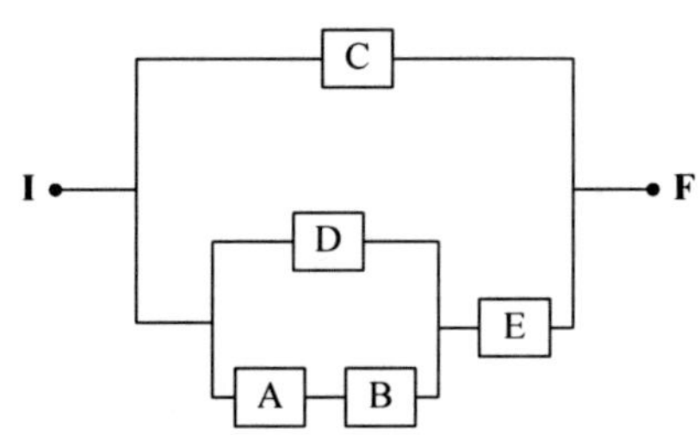

Fig. 32

Fig. 33

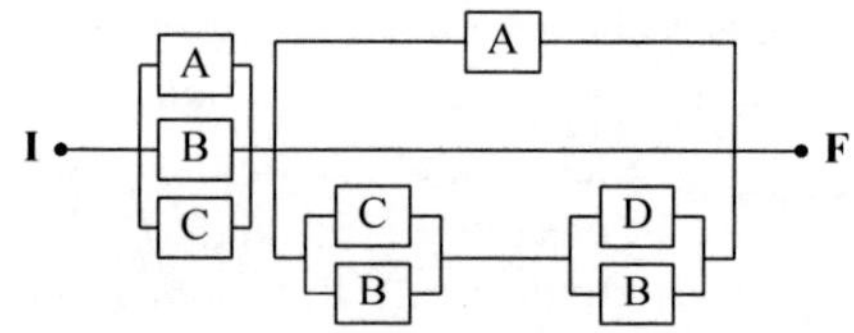

For the network shown in Fig. 33. we obtain the following function of the structure:

$$\begin{aligned}\varphi(a,b,c,d) &= (a+b+c)\cdot(a+(b+c)\cdot(b+d))\\ &= (a+b+c)(a+b+b\cdot d+c\cdot b+c\cdot d)\end{aligned}$$

By the property of absorption

$$b+b\cdot d+b\cdot c=b,$$

so

$$\varphi(a,b,c,d)=a+a\cdot b+a\cdot c\cdot d+a\cdot b+b+b\cdot c\cdot d+a\cdot c+b\cdot c+c\cdot d.$$

Finally, by absorption and using $a^2=a$ and $b^2=b$, we have

$$\varphi(a,b,c,d)=a+b+c\cdot d.$$

Lastly, we can consider one of the most complex networks studied in Fig. 34. It has a structural function as follows:

$$\begin{aligned}\varphi(a,b,c,d,e,j,g,h) &= (a+bd+c)(c+(e+d+j)(g+h))\\ &= (a+bd+c)(c+eg+eh+dg+dh+jg+jh)\\ &= ac+aeg+aeh+adg+adh+agj+ajh+bcd+bdeg\\ &\quad +bdeh+bdg+bdh+bdgj+bdjh+c+ceg+ceh+cdg\\ &\quad +cdh+cgj+cjh\\ &= c+adg+adh+aeg+aeh+agj+ajh+bdg+bdh\end{aligned}$$

Once we know how to obtain a structural function from a representative network of the diagram of a play, we are ready to consider how to obtain the "minimum ties or paths" and also the "minimum lengths".

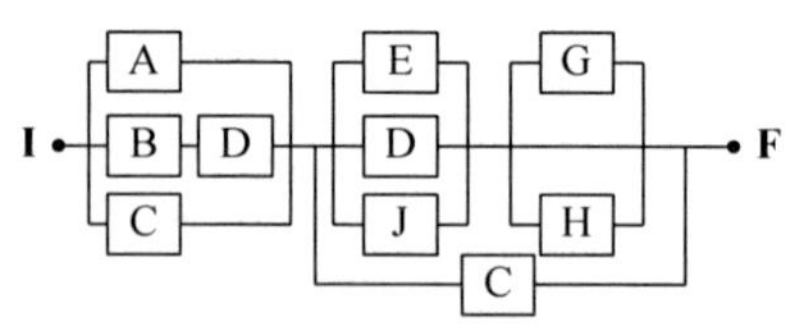

Fig. 34

9 Obtaining Minimum Schemes and Cuts from the Functional Structure

Let us begin with the definition of the "tie" or "path". A path or tie is formed by a subset of athletes such that if they adequately fulfill the desired offensive or defensive functions can make the game scheme work correctly. The adequate performance is evidently to overcome the obstacles that rival athletes introduce. When the diagram of the play is represented by a network, the scheme or path constitutes the path that the ball can take to circulate from point I to point F.

Based on the structural function φ, one can know with certainty if a subset of players forms a tie or path. As such, it is sufficient to give the value 1 to each Boolean variable that corresponds to the athletes considered. If the function takes the value of 1, what we have is a tie. Let us consider the last network as an example.

Recall that the structural function was as follows:

$$\varphi(a, b, c, d, e, j, g, h) = c + adg + adh + aeg + aeh + agj + ajh + bdg + bdh$$

It is proven that $\{B, D, J, H\}$ forms a scheme.

In effect, if we assign a value 1 to the Boolean variables b, d, j, and h, the value $\varphi(x)$ is also equal to 1, although the others are equal to 0:

$$\varphi(a, b, c, d, e, j, g, h) = 0 + 0 + 0 + 0 + 0 + 0 + 0 + 0 + 1 = 1.$$

The same happens with the set $\{C, D, G\}$ or the set $\{A, C\}$, for example.

In contrast, neither $\{A, B, D\}$ nor $\{E, D, G\}$ gives the value 1 for the function φ, as can be proven immediately.

We now arrive at the concept of the "minimum tie" or "minimum path". When a scheme or path does not contain players who form another scheme with reduced numbers, it is said that this scheme or path is minimal. In other words, if a play can be carried out from beginning to end with some athletes, others do not have to intervene.

All of the minimum schemes or paths can be determined without error or omission from the structural function. Therefore, it is sufficient to separate all of the combined Boolean sums. The components of each one form a minimum scheme or path.

The three previous structural functions can serve as an example and proof. Thus, the elements of the combination $\varphi(a, b, c, d, e) = c + de + abe$ are $\{C\}$, $\{D, E\}$, $\{A, B, E\}$, which coincide with those that were visually identified at the beginning of this paper.

The elements of the combination $\varphi(a, b, c, d) = a+b+cd$ are $\{A\}$, $\{B\}$, $\{C, D\}$; these results are also coincident with those that we arrived at visually.

Finally, the nine elements of the addends of the structural function

$$\varphi(a, b, c, d, e, j, g, h) = c + adg + adh + aeg + aeh + agj + ajh$$

correspond to the nine figures that were previously presented based on a visual inspection.

We now move on to the notion of the "cut". A cut is formed by a group of players who impede the development of the play from end to beginning (or theoretically from beginning to end). In reality, it is a line of contention that has the purpose of preventing realization of the play of the opposing team.

Using a structural function, one can determine if a subset of athletes does or does not form a cut. To do this, it is sufficient to substitute the Boolean variables in the function corresponding to the players in question with zeros. If the value of the function is null, we have a cut. We see this with one of the previous structural functions:

$$\varphi(a, b, c, d, e) = c + de + abe.$$

Here, the subset of players $\{A, C, E\}$ creates a cut, and we know this because in giving the Boolean variables a, c, and e the value of 0, we also create a null function:

$$\varphi(a, b, c, d, e) = 0 + d \cdot 0 + 0 \cdot b \cdot 0 = 0$$

The following is not, for example, a null function: $\{A, B, D, E\}$. This is because if the Boolean variable C is equal to one, then

$$\varphi(a, b, e, d, e) = 1 + 0 \cdot 0 + 0 \cdot 0 \cdot 0 = 1.$$

This connects to an important concept, that of "minimum length". By analogy, we will define "minimum length" as that subset of players that has impeded the movement of the ball if you cannot do without any of them. In other words, a minimum length should not include athletes who would not be necessary in a numerically reduced cut.

It is possible to find all of the minimum lengths in a network without error or omission. For this purpose, we suggest transforming the structural function of the network φ into its dual φ^* because the minimum "ties" or "paths" to φ^* are the minimum "cuts" of φ. This represents the players (elements) corresponding to the Boolean variables of the sums of φ^* of the minimum lengths of φ.

To find the dual function of φ^* of a structural function φ, it is sufficient to substitute $(\cdot)$ for $(+)$ and vice versa.

Let us look at some of the structural functions already presented. We can begin with the following:

$$\begin{aligned}
\varphi(a, b, c, d, e) &= c + de + abe \\
\varphi^*(a, b, c, d, e) &= c(d + e)(a + b + e) \\
&= c(ad + bd + de + ae + be + e) \\
&= acd + bcd + ce
\end{aligned}$$

Because the minimum schemes or paths of φ^* are

$$\{C, E\}, \{A, C, D\}, \{B, C, D\},$$

they will also be the minimum "cuts" of the structural function φ. It is observed that they coincide with the findings that we ascertained visually for the figures at the beginning of this paper.

Let us consider the following:

$$\begin{aligned}\varphi(a, b, c, d) &= a + b + cd \\ \varphi^*(a, b, e, d) &= ab \cdot (c + d) \\ &= abc + abd\end{aligned}$$

The minimum "ties" or "paths" φ^* that coincide with the minimum "cuts" of φ are $\{A, B, C\}, \{A, B, D\}$. Note that they coincide with those visually obtained.

We have described construction of the structural function of the representative network of the diagram of play and its dual function with the fundamental elements in the development of a theory of the game systems. Although these are not the only significant considerations, the concept of the minimum tie or path and that of the minimum length are important. We say this because the objective of the game system is perhaps not to arrive at the goal as quickly as possible and using the least possible number of players. The game system in soccer based on "touch", "touch" and more "touch" is enough to support this position. Nevertheless, this does not in any way invalidate the "general theory" that we intend to develop. In these assumptions, other concepts are substituted by minimum bonds/ties or paths and perhaps, although this is not fully established, that of the minimum length. Not to deviate too much from the methodological lines that we have traced here, we do not consider these alternatives here.

References

Adamek, J., Herrlich, H., & Strecker, G. E. (1990). *Abstract and concrete categories*. New York: Wiley.

Beer, G. (1993). *Topologies on closed and closed convex sets*. Dordrecht: Kluwer Academic Publishers.

Brümmer, C. G. L. (1984). Topological categories. *Topology and Its Applications, 18*, 27–41.

Daowen, Q. (2004). Fuzzifying topological linear spaces. *Fuzzy Sets and Systems, 147*, 249–272.

Das, P. (1999). A fuzzy topology associated with a fuzzy finite state machine. *Fuzzy Sets and Systems, 105*, 469–479.

Ertürk, R. (1995). Some results on fuzzy compacts spaces. *Fuzzy Sets and Systems, 70*, 107–112.

García Maynez, A. (1971). *Introducción a la topología de conjuntos*. Ed. Trillas México D. F.

Gil Aluja, J. (1999). *Elements for a theory of decisión in uncertainty*. London, Boston, Dordrecht: Kluwer Academic Publishers.

Gil Aluja, J. (2001). *La pretopología en l gestión de la incertidumbre*. Discurso de investiduracomo Doctor "Honoris Causa" por la Universidad de León. Publ. Universidad de León.

Gil Lafuente, J. (2002). *Algoritmos para la excelencia. Claves para el éxito en la gestión deportiva*. Vigo: Milladoiro.

Giles, J. R. (1987). *Introduction to the análisis of metric*. Cambridge: Cambridge University Press.

Höhle, U. (2001). *Many valued topology and its applications*. Dordrecht: Kluwer Academic Publisher.

Höhle, U., & Sostek, A. (1999). Axiomatic foundations of fixed-basis fuzzy topology. In U. Höhle & S. E. Rodabaugh (Eds.), *Mathematics of fuzzy sets: Logic topology, and measure theory*. Handbook of fuzzy sets series: Vol. 3 (pp. 123–272). Dordrecht: Kluwer Academic Publishers.

Huttenlocher, D. P., Klanderman, G. A., & Rucklidge, W. J. (1993). Comparing images using the Hausdorff distance. *IEEE Transactions on Pattern Analysis and Machine Intelligence, 15*, 850–863.

Jameson, G. J. O. (1975). *Topology and normed spaces*. Londres: Chapman and Hall.

Kaufmann, A. (1983). Prétopologie ordinaire et prétopologie floue. *Note de Travail 115*. La Tronche.

Kaufmann, A., & Gil Aluja, J. (1991). Selection of affinities by means of fuzzy relations and Galois latices. *Proceedings dek XI Euro O.R. Congress*, Aachen, July 1991.

Khedr, E. H., Zeyada, F. M., & Sayed, O. R. (2001). On separation axioms in fuzzifying topology. *Fuzzy Sets and Systems, 119*, 439–458.

Kisielewicz, M. (1991). *Differential inclusions and optimal control*. Dordrecht: Kluwer Academic Publishers.

Kuratowski, K. (1972). *Introducción a la teoría de los subconjuntos borrosos y a la topología*. Barcelona: Vicens Vives.

Lowen, R. (1978). A comparison of different compactness notions in fuzzy topological spaces. *Journal of Mathematical Analysis and Applications, 64*, 446–454.

Malik, D. S., & Mordeson, J. M. (2000). *Fuzzy discrete structures*. New York: Physica-Verlag.

Martin, H. W. (1980). Weakly induced fuzzy topological spaces. *Journal of Mathematical Analysis and Applications, 78*, 634–639.

Menger, K. (1942). Statistical metrics. *Proceedings of the National Academy of Sciences, 28*, 535–537.

Munroe, M. E. (1953). *Introduction to measure and integration*. Reading, MA: Addison-Wesley.

Ponsard, C. (1969). *Un modèle topologique d'équilibre économique*. Paris: Dunod.

Radabaugh, S. E. (1980). The Hausdorff separation axiom for fuzzy topological spaces. *Topology and Its Applications, 11*, 319–334.

Sapena, A. (2001). A contribution to the study of fuzzy metric spaces. *Applied General Topology, 2*, 63–76.

Sugeno, M. (1977). Fuzzy measures and fuzzy integrals, a survey. In M. M. Gupta, G. N. Saridis, & B. R. Gaines (Eds.), *Fuzzy autómata and processes*. Amsterdam: North-Holland.

Sutherland, W. A. (1975). *Introduction to metric and topological spaces*. Oxford: Oxford Science Publications.

Veeramani, P. (2001). Best approximation in fuzzy metric spaces. *Journal of Fuzzy Mathematics, 9*, 75–80.

Xiao, J. Z., & Zhu, Z. H. (2002). On linearly topological structure and property of fuzzy normed linear space. *Fuzzy Sets and Systems, 125*(2), 153–161.

Xiao, J. Z., & Zhu, X.-H. (2004). Topological degree theory and fixed point theorems in fuzzy normed space. *Fuzzy Sets and Systems, 147*, 437–452.

Johnstone, P. T. (1982). *Stone spaces*. Cambridge: Cambridge University Press.

Ying, M. S. (1991). A new approach to fuzzy topology (I). *Fuzzy Sets and Systems, 39*(3), 303–321.

Zadeh, L. (1965). Fuzzy sets. *Information and Control, 8*(3), 338–353.

Zhu, Z. H. (2002). On linearly topological structure and property of fuzzy normed linear space. *Fuzzy Sets and Systems, 125*(2), 153–161.

Zimmermann, H. J. (1978). Results of empirical studies in fuzzy sets theory. In G. J. Klir (Ed.), *Applied general systems research*. New York: Plenum Press.

The Impact of the Global Financial Crisis on Sport in North America

Brad R. Humphreys

Abstract Beginning in early 2007 the global economy entered into a tumultuous period of contraction in real economic activity and disruption in financial markets. This paper explores the effect of the recession and financial crisis that began in mid 2007 on professional sports leagues in North America. While attendance and franchise values declined slightly, and a few teams experienced financial problems, the nature of the sport product and institutional factors associated with the sports industry have, so far, insulted professional sport from significant negative impacts. Structural change in lending may have an impact on new facility construction in the long run. The increasing reliance on revenues from businesses in the form of premium seats, luxury suites, and sponsorship may lead to future problems if the downturn continues for a prolonged period.

1 Introduction

The global economy experienced an extraordinary sequence of economic and financial problems beginning in early 2007. Economies around the globe contracted as real and financial activity simultaneously declined. The size and duration of this contraction has not been matched since the early part of the twentieth century. The economic and financial crisis affected all sectors of the economy; employment, household income, industrial production, and retail sales have all declined significantly since the business cycle peak in mid 2007. In addition, financial institutions failed at alarming rates, and the prices of securities traded on stock markets fell sharply world wide. The impact of these economic and financial problems on certain sectors of the economy, like the labor market, commercial banking, central banking, and stock markets, has been widely documented and analyzed. Here, I assess the impact of the recent economic and financial turmoil on professional sports leagues in North America.

B.R. Humphreys (✉)
Department of Economics, University of Alberta, Edmonton, Alberta Canada T6G 2H4,
e-mail: brad.humphreys@ualberta.ca

S. Butenko et al. (eds.), *Optimal Strategies in Sports Economics and Management*,
DOI 10.1007/978-3-642-13205-6_3,

Sports leagues have a different organizational form, operating methods, goals, and institutional characteristics than firms in other sectors of the economy, so there is good reason to expect them to react differently to economic and financial downturns than other firms. North American professional sports leagues also differ from European sports leagues in a number of important ways that also contribute to a different reaction to economic turmoil. I begin with a discussion of the unique economic characteristics of professional sport, and then describe the economic and financial characteristics of professional sports leagues in North America. After documenting the size and scope of the economic and financial crisis, I explore the ramifications for teams in the National Football League, National Basketball Association, National Hockey League, and Major League Baseball, and briefly touch on the impact on Major League Soccer.

The evidence suggests that attendance, a key source of revenue in professional sports leagues, responded to changes in the business cycle in the past, implying that revenues from ticket sales and other game day revenue streams like concessions and parking may decline. Anecdotal evidence indicates sales of premium seats and luxury suites may decline more because the businesses who buy these relatively expensive products have experienced sharp declines in revenues. Television viewing, another key source of revenues for professional sports leagues, has not declined during the downturn, and most of the revenues earned from broadcast rights come from long term contracts that will not expire until 2012 or beyond; these revenue streams are unlikely to decline in the short or medium term. The disruption of credit flows due to the financial crisis may affect future sports facility renovation and construction projects, which require access to capital markets to complete.

2 The Professional Sports Industry in Context

2.1 Determinants of Demand for Professional Sports

In economic terms, sporting events are entertainment goods that are produced by teams and leagues and purchased by households and businesses. Borland & Macdonald (2003) define the product produced by sports teams as games or contests between two teams and the product produced by leagues as the annual regular season and post season competitions for the league championship. Borland & Macdonald (2003) identify two types of demand for sport: direct demand based on live attendance at sporting events and derived demand based on mediated viewing of sports and the purchase of related goods like merchandise bearing team names or logos.

Borland & Macdonald (2003) also discuss four different factors affecting demand for sporting events: consumer preferences, economic factors, quality of viewing, and the nature of the contest or event. Consumer preferences affect demand for all goods and services in a fundamental way. Consumer preferences for sporting events, however, are in many ways more complex than preferences for other goods or services, because, unlike preferences about consumer necessities like housing or food, consumers' preferences for sporting events depend on phenomena like habit formation, conspicuous consumption, and bandwagon effects that may alter preferences over time. The economic factors affecting demand for sport are similar to those affecting demand for other goods and services. The price of attending or watching a sporting event, the price of substitute activities, the opportunity cost of time, income, and macroeconomic factors like the unemployment rate all affect direct and indirect demand for sporting events in much the same way that they affect demand for television sets or mp3 files. The quality of viewing sports depends on environmental factors like weather, temporal factors like the day and time of the sporting event, as well as the facility and the amenities at the facility (sight lines, concession stands, scoreboards, etc.). The nature of the contest or event depends on the relative quality of the two teams involved and uncertainty about the outcome of the event as perceived by buyers. Generally, contests with greater uncertainty of outcome will generate greater direct and induced demand than contests with a lesser degree of uncertainty of outcome, at both the match and season level.

Many of the factors that affect direct and induced demand for sporting events will not be affected much by the current economic and financial crisis. For example, the relative quality of teams, environmental factors like the weather, and temporal factors related to day and time of games played should all be relatively insensitive to the economic and financial climate in the general economy. In addition, uncertainty about the outcome of games or seasons, identified as an important component of direct and induced demand for sporting events, should be entirely unaffected by general economic and financial conditions in the economy. Much of the core of the sports product, and many of the factors that affect direct and induced demand for sports events, should lie outside the influence of the business cycle and any turmoil in financial markets.

However, the global financial and economic crisis has significantly affected consumers' employment and income, and businesses' revenues. Both consumers and businesses purchase tickets to sporting events, and consumers watch sporting events on television, generating important revenue streams for professional sports teams. Thus the current economic and financial crisis may have a significant impact on the revenues earned by professional sports teams. The next section describes the institutional characteristics of the professional sports industry in North America to explain how the financial condition of teams depend on revenues from consumers and businesses.

2.2 *North American Professional Sports Leagues*

North American professional sports leagues differ from European leagues in a number of important ways, including the revenue sources, roster sizes, league composition, championship determination, and the methods of allocating players to teams and player compensation. In order to understand the effect of the financial and economic crisis on professional sports in North America, it is important to understand the differences in leagues and league structure.

2.2.1 League Structure

There are five major professional sports leagues in North America: Major League Baseball (MLB), the National Football League (NFL), the National Basketball Association (NBA), the National Hockey League (NHL), and Major League Soccer (MLS). Although some might argue that MLS does not constitute a "major" professional sports league, I include MLS in this description because it is comparable to the most popular European professional sports leagues in terms of the sport played, the sport called soccer in North America and football in the rest of the world. There are also a large number of minor professional leagues in North America in each of these sports, as well as in other sports like lacrosse. I restrict my analysis to only these professional leagues because of a lack of financial and economic data from other leagues and because of the prominence of the NFL, MLB, NHL, and NBA in North America.

Professional sports leagues in North America are closed, static leagues. No system of promotion and relegation exists, so the composition of leagues does not change from season to season. The only change in the composition of the five major professional sports leagues occurs when the leagues periodically expand their membership. The NBA last expanded in 1995, adding two teams; the NFL last expanded in 1999, adding two teams; the NHL last expanded in 2000, adding two teams; MLB last expanded in 1998, adding two teams; MLS is the only league currently expanding, adding two teams in 2005, one team in 2007, and one team in 2009, and other new teams will be added in 2010.

All professional sports leagues in North America feature exclusive territorial agreements that divide the US and Canada up into distinct geographic areas where each team enjoys monopoly power, except in a few cases where two teams play in the same large metropolitan area like New York City, Chicago, Los Angeles and the San Francisco Bay area. This system exists because leagues have implicit or explicit exemptions from anti-trust law and can operate as monopolies, completely controlling the number of teams in each league and the location of each team.

2.2.2 Governance and Team Ownership

Although each professional league in North America has a commissioner with nominal power to make league policy, the true power resides with the owners of the teams in each league. The league commissioners primarily handle the day-to-day operation of the league, enforce existing rules governing play, and act as a figurehead and spokesman. No multinational governing body like UEFA exists in North America, and individual teams have much more power than their European counterparts. This stems primarily from the static nature of North American leagues.

With very few exceptions, professional sports teams in North America are privately held businesses that do not issue shares on stock exchanges. None are operated by clubs or organizations of supporters, as is common in Europe. Because of this ownership structure, information about the financial condition of professional sports teams in North America is nonexistent. They are not obligated to release any audited financial data to the public, and seldom open their books for inspection by outside parties. This lack of information makes it very difficult to assess the financial condition of these teams, including the effect of the business cycle on the financial health of teams.

The one exception in North America is MLS, which operates under a single-entity league structure. In MLS, all teams are centrally controlled by the league, revenues are shared across teams in the league, and players ultimately contract with the league, not with individual teams. However, MLS is also a privately held business and does not release financial information to the public.

On the player side, professional athletes in team sports in North America are all unionized. The National Football League Players' Association, the National Basketball Players' Association, the Major League Baseball Players Association, the National Hockey League Players' Association, and the major League Soccer Players Union represent players in the main professional leagues in North America. In all these leagues, a collective bargaining agreement (CBA) between the players' union and the league governs all aspects of the economic interaction between players and teams. CBAs typically last for 5 years and the re-negotiation of a CBA often leads to significant work stoppages in North American professional sports leagues. For example, the NHL lost the entire 2004–2005 season to a work stoppage, teams in the NBA played only 50 games, instead of the usual 82, in the 1998–1999 season because of a work stoppage, and MLB canceled the last 2 months of the regular season and the entire post-season, including the World Series, in 1994 because of a work stoppage.

2.2.3 Revenues

Teams in North American sports leagues earn revenues from five broad sources: gate revenues from ticket sales, facility-based revenues from concessions and other game related activities like parking, local and national broadcast rights fees, and licensed merchandise sales revenues. Facility-based revenues depend primarily on attendance at games, since the more people attending games, the more money spent on these goods and services. Transfer fees do not exist in North American sports leagues, and some leagues prohibit that sale of players for cash. The relative contribution of each type of revenue differs across leagues, but in general, ticket sales and revenues related to attending games represent the largest source of revenues for professional sports teams in North America; media rights fees, including television and radio broadcast rights, represent the second largest source of revenues.

For most of the economic history of professional sports in North America, sales of tickets to individual fans, either in the form of "season ticket" packages to all home games, or tickets to individual games, represented the single largest source of revenues to professional sports teams. Beginning in the 1990s, teams began to generate significant revenues from premium seating, including luxury boxes and other "club seats" located in special sections of sports facilities separated from general admission seats (Mason & Howard, 2008). A new stadium or arena in North America typically contains between 70 and 140 luxury suites and several thousand premium seats. Tickets for seats in these premium locations typically cost two to four times more than other tickets and come with enhanced concessions and other amenities like personal television screens. Luxury suites must be leased for an entire season at an average price often exceeding $100,000, not including mandatory purchases of food and beverages at each game. The primary customers for premium seats and luxury suites are not individual fans; these seats are primarily sold to corporations and other businesses (Mason & Howard, 2008).

Another important source of revenues in professional sports in North America is sponsorship. In North American leagues, sponsorship revenue comes from facility naming rights fees, revenues from signage in sports facilities, and rights fees associated with product endorsements. Facility naming rights deals can be quite lucrative in North America. For example Reliant Energy paid $10 million per year for the naming rights to the stadium where the Houston Texans play football and Royal Phillips Electronics paid $9.3 million per year for the naming rights to the arena where the Atlanta Thrashers HNL franchise and the Hawks NBA franchises play. Stadium signage rights fees are typically smaller. Product endorsement revenues come from sponsorship deals that make certain products the "official" product of some sports team or league. There are no jersey sponsorship deals in the NFL, NBA, MLB or NHL. Corporate advertising on jerseys is forbidden in these leagues at this time, unlike the current practice in Europe. O'Reilly and Nadeau (2006) document an increase in revenues from sponsorship in North American sports leagues over the past few decades.

2.2.4 Costs

Like professional sports teams all over the world, player compensation accounts for the largest operating expense incurred by teams in North American sports leagues. One unique feature of the payroll costs of North American sports leagues is the presence of salary caps that limit the total compensation paid to players in the league. The current collective bargaining agreements between players' unions and the NFL, NBA, NHL, and MLS include salary caps or other restrictions on player salaries. Only MLB does not have a formal salary cap in place at this time. Salary caps in North American sports leagues limit total payroll on each team to a specific fraction of total revenues earned by teams in each season. New players are acquired either through amateur entry drafts that assign rights of new players to specific teams and limit salaries paid to these new players, through trades, or through free agent signings. There are almost no transfer fees in North American professional sports leagues. Players change teams through trades or free agency. The only exceptions are that professional baseball players moving from Japan to North America sometimes require the payment of a transfer fee to a Japanese team, and professional hockey players moving from European teams to NHL teams must require a $200,000 transfer fee be paid to the European club.

Professional athletes in North American leagues have limited free agency. In general, new players in leagues are assigned to teams through amateur entry drafts and the teams assigned the rights to these players can keep them under contract for a number of seasons before players can become free agents and sell their services to the highest bidder on the open market. MLB players can become free agents after playing 6 seasons, NBA players can become free agents after playing 3 or 4 seasons, NFL players can become free agents after playing 3 seasons, and NHL players can become free agents after playing seven seasons or turning 27 years of age. In addition, long term contracts, contracts of duration 3 or more years, are relatively common in North American Sports Leagues.

The existence of salary caps, limited free agency, and long-term contracts mean that payroll expenses are relatively stable in North American professional sports leagues. Salary caps place limits on the number of new players that can be acquired in any season, and the limited nature of free agency means that players with relatively little experience are paid low salaries. It also means that teams in financial difficulty cannot raise large sums of money quickly through transfer fees earned by selling off star players.

2.2.5 Summary

The structure of the professional sports industry in North America provides some insulation from the business cycle and other financial crises, especially on the cost side. However, professional sports teams rely heavily on consumer spending on tickets and other facility-based revenues like concessions. Media rights contracts, which ultimately depend on consumers' viewing habits, also play an important role in revenues generated by teams. In the last decade, professional sports teams have also increased their reliance on spending by businesses on premium seating, luxury seating, and sponsorship. Since both consumers and businesses have been profoundly affected by the current global financial and economic crises, teams may experience declines in revenues as a result of the crisis. The next section documents the scope of the global financial and economic crisis and assesses the impact of the crisis on the finances of professional sports teams in North America.

3 The Dimensions of the Financial and Economic Crisis in North America

The economies of the United States and Canada are in the midst of a significant economic downturn which began in 2007. This downturn can be described as a recession, a macroeconomic event characterized by a general downturn in economic activity over a sustained period of time. Recessions feature co-movement of most or all aggregate economic indicators, including employment, industrial production, sales, income, and household spending. This downturn affected both the real economy, reducing employment, industrial production, income, and consumer spending, and the financial sector, generating significant interruptions in the flow of capital and the ability of households and firms to borrow against future earnings. The events in the real economy and the financial sector are related, although the timing and extent of the downturn differ somewhat in these two areas. The decline in real economic activity in past recessions has been temporary—aggregate production, income, sales and employment have returned to their pre-recession long-run trends eventually, and sometimes rapidly. The financial consequences of this downturn may, on the other hand, feature important long-term permanent structural changes (Chari, Christiano, & Kehow, 2008). Although the worst period of contraction appears to be over, economic conditions have not yet begun to improve, and many sectors of the economy may not recover for quite some time. Whenever this downturn ends, it will probably be the longest and deepest recession in the United States since World War II (Kliesen, 2009).

The decline in real economic activity, like production and earnings, began in December 2007, when the NBER Business Cycle Dating Committee determined that the economy in the United States experienced a turning point, moving from expansion to recession. The NBER dating committee determines the precise point in time when the economy moved from expansion to contraction after a long period of deliberation and analysis. The December 2007 business cycle peak was announced on December 1, 2008.

Figure 1 documents the decline in several key economic indicators for the US economy after the business cycle turning point in December 2007. Retail sales, defined as revenues earned by establishments that sell goods and services to households, employment, defined as the number of persons working, and industrial production, defined as the dollar value of goods produced, are considered important indicators of macroeconomic performance in developed economies. On Fig. 1, the estimated monthly value of each of these three economic indicators was normalized to 100 in December 2007, which is represented by zero on the horizontal axis. Negative numbers on the horizontal axis refer to months before December 2007, and positive values to months after December 2007. Since the values of each variable

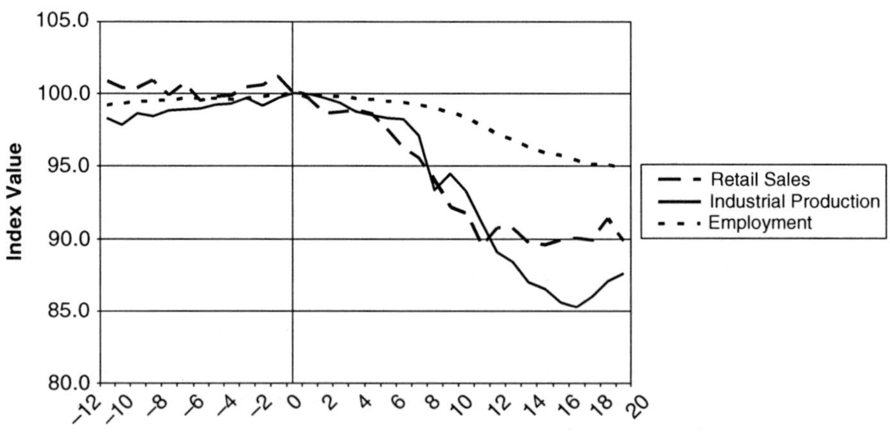

Fig. 1 Decline of key US economic indicators

have been normalized to 100, the percent decline in each variable can easily be read off of the vertical axis.

From Fig. 1, retail sales, industrial production and employment all declined in the months following the December 2007 peak in the business cycle. The decline was small in the first few months, but picked up considerably in the summer of 2008, roughly 6 months after the peak. Industrial production and retail sales fell by more than employment in percentage terms. By the end of 2008 retail sales and industrial production had declined by 10% and employment by almost 5%. However, employment continued to decline 2 years after the peak, while retail sales and industrial production stabilized somewhat.

The decline in retail sales and industrial production are both important for understanding the effect of the current economic and financial crisis on professional sports leagues in North America. As discussed above, professional teams in North America have moved from a heavy reliance on consumer spending, in the form of single game and season ticket sales, concessions and other game day spending, and licensed merchandise for revenues to a much larger reliance on business spending for revenues. This business spending takes the form of premium seat and luxury box leases, facility naming rights, and other sponsorship revenues. The decline in retail sales and industrial production means that business revenues have fallen sharply since the start of the economic and financial crisis in 2007. Businesses have considerable flexibility in adjusting their expenditures, and as profit maximizing organizations, businesses will rapidly cut costs when faced with declining revenues in order to avoid losses. Businesses cut costs by laying off employees, and reducing expenses on other inputs to production. Economic theory predicts that profit maximizing firms will reduce costs based on the marginal revenue product generated by each input. Spending by businesses on sports, in the form of premium seats, luxury suites, and sponsorship deals, is indirectly related to the production of output in most cases. The marginal revenue product of inputs indirectly related to production may be relatively low, implying that businesses will reduce spending on these inputs first. Businesses may drastically reduce spending on sports related inputs to production during an economic downturn.

These three economic indicators tell only part of the story of the decline in real economic activity after the December 2007 business cycle peak. The decline in employment also reflects significant declines in earnings for the households of unemployed workers. Since recessions feature co-movement of most economic variables, many additional economic variables also declined at about the same time, and by similar magnitudes.

The timing of the financial crisis is harder to determine exactly and the impact both wider and harder to assess. The first signal of an impending financial crisis in the US came in February, 2007, when the Federal Home Loan Mortgage Corporation ("Freddie Mac") announced that it would stop buying the most risky type of home mortgages, those falling in the "subprime" category, as well as any mortgage-related securities based on derivatives of these mortgages. Subprime mortgages refer to loans made to individuals with poor credit histories to purchase a home. In April 2007, New Century Financial Corporation, a large lender in the sub prime mortgage market, declared bankruptcy. By the end of 2007, the subprime mortgage problem had spread well beyond this relatively small area in the financial sector. Large financial institutions in the US and Europe began reporting large losses and, in short order, a credit crunch developed that featured reduced liquidity across the financial sector and limits on the flow of credit between financial institutions (Mizen, 2008).

The event that touched off the financial crisis, a sharp increase in defaults on subprime mortgages, was relatively small. Blanchard (2009) estimates that the value of losses on subprime loans and derivative securities based on subprime loans, was only $250 billion in the US as of October 2007. Unfortunately, structural factors in financial and housing markets rapidly amplified the subprime loan collapse. Blanchard (2009) estimates the worldwide loss in all financial markets (stocks, bonds, real estate) to be more than $26 trillion, roughly 100 times the initial subprime mortgage losses. In addition, the financial crisis featured a dramatic decline in the flow of capital between lenders and from lenders to borrowers. Brunnermeier (2009) characterizes this as a liquidity squeeze where lenders fear for their future access to capital markets and begin hoarding funds.

The mechanisms through which the relatively small subprime losses were amplified are beyond the scope of this paper. The key is to understand the ramifications of the financial crisis for professional sports teams in North America. For North American professional sports teams, the key consequence of the financial crisis will be the extent to which access to borrowing, in the form of short term commercial paper, and long term debt to finance new facility construction and renovation, will be reduced. All businesses require some short term borrowing to smooth out temporal differences between revenues and expenses. For example, broadcast rights fees are paid out in lump sums and game day revenues are realized only during the season while salary expenses are monthly or bi-weekly; these temporal differences may require short term borrowing to meet payroll. Professional sports teams also must periodically replace or renovate stadiums and arenas and practice facilities. These expensive capital investment projects require borrowing against future revenues. The financial crisis will clearly bring about significant structural changes in the flow of capital from savers to lenders to borrowers. Problems for professional sports teams will emerge in the long run if team's access to short and long term borrowing declines.

4 Impacts of the Financial and Economic Crisis

4.1 Attendance

4.1.1 General Admission

The response of attendance at professional sporting events to past economic downturns can shed light on the effect of the current economic and financial crisis on sports. The exact date of turning points in the business cycle have been determined by the National Bureau of Economic Research (NBER) in the US for each recession and expansion since the 1950s. The exact date of each turning point in the business cycle—a peak when the business cycle switches from expansion to recession and a trough when the business cycle switches from recession back to expansion—has been identified to the month by the NBER as part of the business cycle reference date project undertaken by the Business Cycle Dating Committee. These business cycle reference dates define the start and end of recessions and expansions with great precision in the US over a long period of time. Data on attendance at professional sporting events over long periods of time are, however, difficult to obtain for many professional leagues in North America. After considerable searching, I located data on total league attendance in each season in MLB from 1900 to 2008, in the NBA from 1949 to 2008, in the NHL from 1960 to 2008, and in the NFL from 1970 to 2008. However, total NFL attendance for 1992 and 1998 were not available.

Figure 2 shows total attendance for each season in the NFL, NBA, MLB and NHL as well as the peak of the business cycle, over the past 100 years. Since only peaks of the business cycle are identified, the period between each vertical line on Fig. 2 represents both a recession and the ensuing expansion. The average recession lasted 14 months and the average expansion lasted 44 months over this period. The annual attendance plots show a great deal of variation, in part because of work stoppages and changes in the number of teams in each league and the number of games played in each season. Many of the large declines are associated with work stoppages that resulted in the cancellation of a significant number of games.

A careful examination of Fig. 2 reveals some instances when total attendance appears to decline during recessions. For example, MLB and NFL attendance drops significantly in the early 1980s, when two recessions hit the US economy in January 1980 and July 1981. However, both leagues experienced work stoppages during

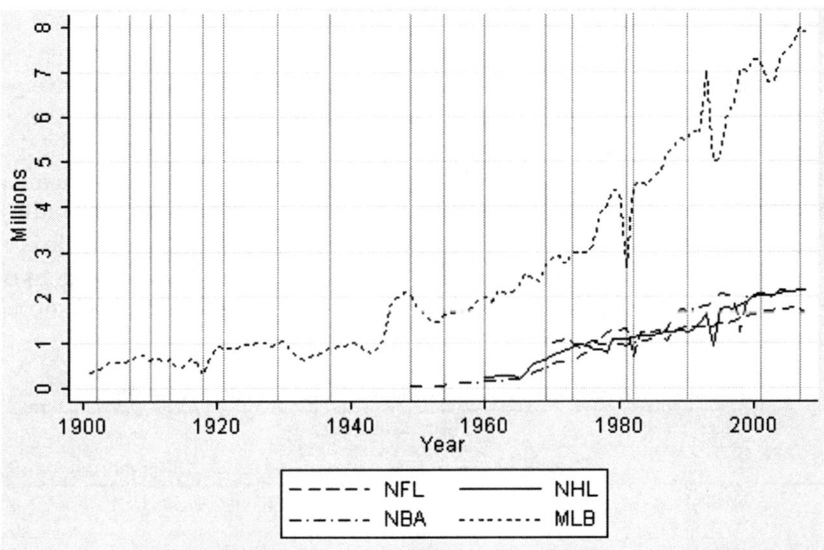

Fig. 2 Total attendance and peaks

this period, which may explain some of the decline in attendance. Also note that, based on Fig. 2, MLB remains the most popular sport among these four in terms of attendance, and all four leagues have experienced significant growth in attendance over time.

Instead of relying on visual examination to understand the effect of recessions on attendance, I estimated a simple linear regression for each league in order to better understand how attendance changes during recessions. The dependant variable in each regression was total attendance in the league in each season; the explanatory variables were a time trend term to capture the secular increase in demand for each sport and a variable that indicated the number of months during each season when the US economy was in a recession, according to the NBER business cycle reference dates. Using the number of months in each season that the economy was in recession is a better indicator of the effect of the business cycle than a simple indicator variable because of the timing of the start of recessions and the different start and end dates of the season in each league. For example, suppose a recession started in March of a given year and ended in November. This recessionary period would include the entire 7 months of the MLB regular season, which runs from the beginning of April until the end of October; it would include 4 months of the NFL regular season, which runs from the beginning of September until early January, and it would include only four of the 6 months in the NBA and NHL regular seasons, which run from early November until mid April.

Table 1 shows the results from this regression analysis: parameter estimates, t-statistics on two-tailed tests of the null hypothesis that the parameter estimate is equal to zero, and R2. The parameter estimate on the time trend variable is the average annual increase in total attendance in each league. Total attendance in MLB was increasing by the most, more than 600,000 per season on average, and total attendance in the NFL was growing by the smallest amount. These parameter estimates are highly significant, with t-Statistics in excess of 10. This result is not surprising given that the NFL focuses much more on its television audience than on live game attendance while MLB focuses on gate attendance and other facility-related revenue sources.

The parameters of interest in Table 1 are on the variables indicating the number of months during each season when the economy was in a recession, as identified by the NBER business cycle reference dates. The average value of this variable is 1.67 in MLB, 0.88 for the NBA, 0.64 for the NFL, and 1.02 for the NHL. Recessions were relatively rare events during the sample period. The parameter estimate is negative

Table 1 Regression results, total annual league attendance model

	Time trend		Months in recession			
Sport	Estimate	T-Stat	Estimate	T-Stat	Observations	R2
NFL	215,847	12.57	–224,122	–1.57	37	0.825
NBA	432,458	32.81	–184,992	–1.58	60	0.950
NHL	411,779	28.10	70,818	0.73	48	0.946
MLB	665,678	23.33	340,049	1.03	109	0.843

for only the NFL and the NBA. In both cases, these parameter estimates are only marginally significant. The P-value on the two-tailed test of the hypothesis that the parameter is equal to zero is rejected at the 12% level for both sports. Given the relatively small sample sizes for these regressions, these parameter estimates provide weak evidence that total attendance in the NFL and NBA decline during recessions, and that the longer the recession lasts, the larger the decline in total attendance. The parameter estimates on the number of months in recession variable are not statistically different from zero in the NHL and MLB, suggesting that total attendance in these leagues does not respond to downturns in the economy. Attendance in the NFL and NBA may be more sensitive to the business cycle than in MLB and the NHL because ticket prices in the NFL and the NBA are, on average, higher. The more expensive the good or service, the more likely are households to reduce spending on that item when budgets get tight, especially for entertainment goods like tickets to professional sporting events, which are not necessities like food, clothing or shelter.

The regression models explain quite a bit of the observed variation in total annual attendance in these four leagues. The regression model explains between 84.3% and 95% of the observed variation in total annual attendance. The F-statistics on the overall significance of the regression models are also large (80.3 for the NFL, 541 for the NBA, 284 for MLB and 395 for the NHL), suggesting that, taken together, the models fit the data well.

Overall, the evidence indicates that the current economic and financial crisis will have a minor impact on gate attendance in these four North American professional sports leagues. Although visual inspection of the time path of annual total attendance uncovers some instances where total attendance appears to have declined during recessions, the regression results find only weak support for the idea that attendance will decline, with one caveat. The current economic and financial crisis appears to be lasting longer, and also seems more severe, than the average recession in the sample period. The US economy probably has not yet switched from recession to expansion—although the NBER business cycle dating committee only determines turning points after considerable time has passed—and the longer the current crisis lasts, the more likely it will affect attendance at professional sporting events.

4.1.2 Premium Seating

No comprehensive statistics are kept on premium seating sales in North American professional sports leagues. As mentioned above, the price of these seats, which include club seats and luxury boxes, are very high and many must be leased for one or more seasons. Premium seats and luxury boxes are primarily purchased by businesses, not individual fans.

In early November 2009, TicketOS, an on-line ticket purchasing company, announced the creation of an index to measure the use of premium seats and luxury boxes at sporting events by corporations in North America, based on a survey of large corporations. According to initial reports, club seat use averaged 73% in October 2009, and luxury box use averaged 67%. Both were down relative to use in October 2008. However, corporate season ticket sales were reported to be up in ten venues located in large metropolitan areas like New York City, Boston, Washington DC, and Atlanta.

Anecdotal evidence also indicates that revenues from premium seats and luxury suites have declined in recent seasons. According to a recent article in the Indiana Business Journal (Schoettle, 2009), 10 luxury suite leases at Conseco Fieldhouse, home of the Indiana Pacers, were unsold just before the start of the NBA season and NFL teams in Cleveland (20 vacant suites), Jacksonville (20 vacant suites), Miami (30 vacant suites), and Detroit (35 vacant suites) were also having difficulty leasing luxury suites.

4.2 Broadcast Revenues and TV Viewing

North American professional sports teams generate broadcast rights revenues from two sources: local broadcasting rights and national broadcasting rights. Teams negotiate local broadcast rights individually with local media providers and these revenues are not shared with the rest of the teams in the league. National broadcast rights contracts are negotiated jointly by the league and these revenues are shared equally by all teams in the league. National broadcast rights contracts extend over multiple seasons and the value of these contracts varies considerably across leagues.

It is difficult to determine the annual value of local broadcast rights contracts systematically for all North American professional sports leagues, so assessing the effect of the current economic and financial downturn on these rights fees is extremely difficult. Clearly, local broadcast rights fees will be related to the size of the market that each team plays in, the quality of the team, the preferences of fans in the local market for viewing sports on television, and the presence and price of substitutes for sports on television in each market. In general, local broadcast rights fees will vary with overall television viewing trends. In both the US and Canada, overall television viewing did not decline following the beginning of the economic downturn in late 2007. Average television hours viewed in Canada remained steady at about 28.5 h per week over the period 2004–2009. In the US, average television hours viewed per week increased from 36 h per week in 2007 to 37.75 h per week in the first quarter of 2009. If these trends reflect sports viewing habits, then local broadcast rights fees are not likely to decline as a result of the economic downturn.

Considerably more information exists about national broadcast rights fees, which are negotiated over long periods of time. Table 2 summarizes the current national broadcast rights fees for the major North American professional sports leagues as of late 2009.

Note the wide disparity in the value of rights fees across leagues. The NFLs five national rights fees contracts generate in excess of $20 billion over the period 2006–2013. At the other end of the spectrum, MLS earns a paltry $8 million per season from ESPN for national broadcast rights. The contract between NBC and the NHL did not involve rights fees. Instead, NBC will pay the NHL the residual revenues after all costs related to broadcasts have been covered. Almost all of these

Table 2 National broadcast rights fees in North America

League	Rights holder	Total broadcast fee	Contract period
NFL	ESPN	$8.8 billion	2006–2013
NFL	Fox	$4.27 billion	2006–2011
NFL	CBS	$3.73 billion	2006–2011
NFL	NBC	$3.6 billion	2006–2011
NFL	DirecTV	$3.5 billion	2006–2010
MLB	Fox/TBS	$3.0 billion	2006–2013
NBA	ABC/ESPN	$2.4 billion	2008–2016
NBA	TBS	$2.2 billion	2008–2016
NHL	Versus	$290 million	2007–2011
NHL	NBC	NA	2007–2011
MLS	ESPN	$64 million	2006–2014

contracts run through 2011, and the one contract up in 2010, between the NFL and satellite TV company DirecTV for a pay-per-view package of all NFL games, is relatively small. Thus the global economic and financial crisis will have little short run impact on revenues earned by professional sports leagues in North America.

If the economic climate has improved by the time these contracts expire, the long run impact may also be minimal. These revenue streams will only decline if the popularity of sports on television declines in the next several years. The ratings for the first week on the NFL season in 2009, a 10.3 share of the total television sets being watched, were up significantly over the ratings for the first week of the 2008 NFL season, an 8.8 share. In MLB, the ratings for the 2009 World Series, played in October, were also up significantly relative to the past 5 years. Some of this increase can be attributed to the presence of the New York Yankees in the World Series, but viewing did not drop, again suggesting that sports television viewing in North America will not be adversely affected by the economic and financial crisis.

4.3 Franchise Values and Bankruptcy

Another method for assessing the effect of the economic and financial crisis on professional sports leagues is to examine changes in the estimated value of franchises. Forbes Magazine produces annual estimates of the value of every professional sports franchise in North America. As was mentioned above, these franchises are privately held corporations that do not release audited financial data to the public. Because of this lack of information, the Forbes Magazine estimates are based on observable factors like attendance, media rights fees, and payroll along with the use of standard financial multipliers. Humphreys and Mondello (2008) find that these estimates contain considerable noise when compared to actual franchise sales prices. Despite this limitation, the Forbes estimates are the only information available on annual sports franchise values, since no professional sports team in North America currently issues shares that are tradable on any stock exchange. At the time this paper

Table 3 Average change in franchise values 2008–2009

	Average franchise			
Sport	Value increase (%)	Decreasing (%)	Average decrease (%)	Largest decrease (%)
NFL	0.16	25	–3.37	–7
NHL	1.27	47	–4.64	–11
MLB	1.23	33	–5.70	–12

Source: Forbes Magazine

was written, Forbes Magazine had released the 2009 franchise value estimates for the MLB, NFL and NHL.

Table 3 summarizes the Forbes estimates of changes in franchise values from the 2008 season to the 2009 season. Note that this period coincides with the worst of the economic and financial crisis in North America, based on the discussion above. If teams feel the pinch from the ongoing recession and financial disruption, it should be reflected in these estimates. All three leagues saw extremely modest increases in average franchise values from 2008 to 2009. These increases are not inflation adjusted, but the inflation rate in the US, based on changes in the Consumer Price Index, was essentially zero over this period, so deflating would have no impact on the estimates of increases in franchise values.

From Table 3, eight of the 32 NFL franchises declined in value over the past year, and the average decline among these eight was about 3.3%. Almost half of the NHL franchises declined in value, and one third of the NBA franchises declined in value. The size of the decline in the NBA and NHL were larger than in the NFL, and a handful of hockey and basketball franchises saw double digit declines in their value. These declines in franchise values suggest that professional sports teams have been adversely affected by the economic and financial crisis. Humphreys and Mondello (2008) report very few instances of declines in the value of professional sports franchises over the past 40 years, and the hedonic franchise price index they develop contains relatively few short run downturns in the average franchise price. However, the decreases in franchise values are relatively small when compared to the decline in the value of publicly traded firms in the US. Broad market indexes in the US like the Standard and Poor's 500 declined by more than 30% from 2008 to 2009, indicating much larger average declines in the value of publicly traded firms, so these estimates of the change in the value of sports franchises suggest that professional sports leagues fared much better than many other sectors in the US economy over the past year.

Another indicator of the effect of the economic and financial crisis on businesses is bankruptcy filings. In the US, any business or individual that is unable to pay off its debt can appeal to a federal bankruptcy court for protection from creditors under a law commonly referred to as Chap. 11. Filing under Chap. 11 bankruptcy implies that the business or individual will be unable to fully pay off its debts, signaling extreme financial duress. Chapter 11 filings involve a reorganization of debt, and most firms undergoing Chap. 11 reorganization continue to operate. The economic and financial crisis has had a clear effect on bankruptcy filings, which increased by 31% from 2007 to 2008.

In the past year, two professional sports teams have filed for bankruptcy under Chap. 11: the Chicago Cubs and the Phoenix Coyotes. Both cases involve special circumstances that make them non representative of the overall financial health of the sports industry. The Chicago Cubs were owned by the Tribune Corporation, a media conglomerate that also owned numerous newspapers and television stations. The Tribune Corporation filed for bankruptcy under Chap. 11 in 2008. The Cubs were not bankrupt in the sense that the team was able to pay its debts. However, the team filed for bankruptcy in order to make the sale of the team possible, and to avoid paying $300 million in taxes, the team filed for bankruptcy. The Cubs were sold to an individual owner, Tom Ricketts, for $875 million in October 2009. Although the Phoenix Coyotes hockey team had allegedly been losing money for several years, the team's bankruptcy was directly caused by the financial problems of the team's owner, Gerry Moyes, who was also involved in extensive real estate speculation. The Coyotes were sold to the National Hockey League for $150 million in October 2009.

5 Conclusions

While the industry has not escaped from the current economic and financial crisis unscathed, the evidence presented above indicates that the overall impact in the short run has been relatively minor. On the revenue side, attendance may decline slightly in the NFL and NBA, but overall attendance is unlikely to decline significantly as a result of the recession. Television viewing has not dropped, and the long-run broadcast rights contracts currently in place insulate the leagues from any decline in broadcast rights revenues for the next several years. Broadcast rights fees could decline when the current contracts run out in 3–4 years, but by that time the general economic environment could improve significantly. The biggest short term consequence on the revenue side is the possibility that revenues from luxury suites, premium seats and sponsorship could decline due to reduced spending by businesses. The increasing reliance of professional sports teams in North America on spending by businesses over the past decade means that these declines could be significant. The general decrease in estimated franchise values documented above probably reflects declines in revenues from businesses more than any other factor. On the cost side, the existence of salary caps and limited free agency insulates payroll costs from the influence of outside economic and financial forces. The biggest threat on the cost side is the expiry of the Collective Bargaining Agreements between the leagues and players' unions in 2011 (MLB, NBA) and 2012 (NFL and NHL).

The evidence from estimates of franchise values suggests a modest impact on teams. Although a large number of teams experienced small declines in estimated value, the impact appears much smaller than in other sectors of the economy. However, the impact of the current economic downturn could be more severe if the current recession, which began in December 2007 and is now 23 months old, continues for a significantly longer period of time. The average length of all recessions in the

US since 1857 was 17 months, and the 10 recessions since 1945 have lasted an average of just 10 months. The current recession is already quite long in historical context. Another year of recession might bring about additional financial problems in the professional sports industry in North America.

In the long run, the biggest potential area for problems in professional sports in North America is the effects of the financial crisis on access to credit. The professional sports industry in North America is in the middle of a stadium and arena construction boom that began in the early 1990s (Zimbalist and Long, 2006). Zimbalist and Long (2006) report that the median cost of new stadiums since 2000 was $455 million, and the median cost of new arenas was $210. Several recent facility construction projects, including new stadiums for the New York Yankees and Dallas Cowboys, cost more than $1 billion. The pace of new sports facility construction does not appear to have slowed in recent years. Although these projects typically involve substantial public funding, teams are increasingly pressured to provide their own, privately generated funds for new facility construction. The global financial crisis will have important effects on borrowing and the operation of credit markets. The extent and nature of these changes are still unknown, but it is reasonable to expect that access to credit will not be enhanced by the new regulations and business practices adopted by lenders. Financing of new facility construction projects, as well as the renovation of existing facilities, will be more difficult in the future. The pace of new facility construction will probably decline, and only franchises with the strongest potential for future revenue generation will be able to borrow to finance facility construction. The effects of restricted access to credit on the core sports product, contests and league championships, are difficult to predict. Quinn, Bursik, Borick, and Raethz (2003) found no relationship between new facilities and on-field success in all four professional sports leagues in North America. But if prolonged economic and financial problems lead to reduced spending on maintenance on existing facilities, the financial and on-field consequences could be quite different.

References

Blanchard, O. (2009). *The crisis: Basic mechanisms, and appropriate policies* (International Monetary Fund Working Paper).

Borland, J., & Macdonald, R. (2003). Demand for sport. *Oxford Review of Economic Policy, 19*(4), 478–502.

Brunnermeier, M. K. (2009). Deciphering the liquidity and credit crunch 2007–2008. *Journal of Economic Perspectives, 23*(1), 77–100.

Chari, V. V., Christiano, L., & Kehow, P. J. (2008). Facts and myths about the financial crisis of 2008. Federal Reserve Bank of Minneapolis working paper #666.

Humphreys, B. R., & Mondello, M. (2008). Determinants of franchise values in North American professional sports leagues: Evidence from a hedonic price model. *International Journal of Sport Finance, 3*(2), 98–105.

Kliesen, K. L. (2009). Dating the end of the recession: Evaluating the economic indicators. *Federal Reserve Bank of St. Louis Economic Synopses, 28*, 1–2.

Mason, D. S., & Howard, D. R. (2008). New revenue streams in professional sports. In B. R. Humphreys & D. R. Howard (Eds.), *The business of sport* (Vol. 1, pp. 125–152). Westport, CT: Praeger.

Mizen, P. (2008). The credit crunch of 2007–2008: A discussion of the background, market reactions, and policy responses. *Federal Reserve Bank of St. Louis Review, 90*(6), 531–568.

O'Reilly, N. J., & Nadeau, J. P. (2006). Revenue generation in professional sport: A diagnostic analysis. *International Journal of Sport Management and Marketing, 1*(4), 311–330.

Quinn, K. G., Bursik, P. B., Borick, C. P., & Raethz, L. (2003). Do new digs mean more wins? *Journal of Sports Economics, 4*(3), 167–182.

Schoettle, A. (2009). Sales of luxury suites slow at most Indianapolis sports venues. *Indiana Business Journal*, 1-2, October 19.

Zimbalist, A. & Long, J. G. (2006). Facility finance: Measurement, trends, and analysis. *International Journal of Sport Finance, 1*(4), 201–211.

Managing and Modeling the Combination of Resources in Professional Sporting Events

Lionel Maltese and Lucien Veran

Abstract With a view towards integrating the Business Model concept into Resource-Based thinking, this research paper questions the concept with respect to its potential for formalizing a firm's resource arrangement and control methods. The clinical study of three cases of international sporting event organization (Roland Garros, BNP Paribas Masters and Open13) enables a comparison of Business Models which rely on the same pool of resources but which structure and manage them differently. The main result identified in the context of an unstable sports environment, is the importance of such organizations' aptitude to deploy and control resources which have been built up with certain stakeholders. Beyond the comparison of different Business Models within a similar environment, an empirical sequential approach for building up and managing a firm's assets is suggested. The analysis of the development of each Business Model also enables us to understand how strategies for limiting the dependence of an organization on the reputation of its main suppliers are rolled out.

1 Introduction

The Business Model (BM) concept has been attracting the attention of the academic community (Zott & Amit, 2007) for some years, and within that community, trends known as resource, competency and capacity-based approaches (Teece, 2007). RBV sees companies as different collections of physical and intangible resources and capabilities, which determine how efficiently, how effectively a company performs its functional activities. The RBV trend of thought emerged in the early 1990s with the founding works of Wernerfelt (1989), Barney (1991) as well as Grant (1991) and has followed different paths starting with a strictly internal approach to the firm based on its intention and desire to transform the competitive environment starting

L. Maltese (✉)
Paul Cezanne University, Aix-Marseille, Marseille, France; Euromed Management, Marseille, France
e-mail: lionel.maltese@univ-cezanne.fr

S. Butenko et al. (eds.), *Optimal Strategies in Sports Economics and Management*, DOI 10.1007/978-3-642-13205-6_4,

with its own local capacities and resources (Hamel & Prahalad, 1990). In the same frame of reference, the purpose of this research is to try to understand the BM concept as one of the possible extensions of the Resource-Based View (RBV) trend of thought and more particularly of thinking regarding the management process of combining strategic assets (Morgan, 2000; Sirmon, Hitt, & Ireland, 2007; Sirmon, Gove, & Hitt, 2009; Bingham & Eisenhardt, 2008) while taking into account the contribution made by environmental players (Rindova & Fombrun, 1999) and the role played by certain stakeholders (Coff, 1999) in order not to limit the analysis to internal issues.

Questions relating to the a firm's arrangement of its assets and their scope of control, which may be shared, to a greater or lesser degree, with certain stakeholders who are contributors in terms of resources, will be addressed by combining studies of clinical cases with action research.

The industry studied is that of professional sporting events, through the cases of the French international tennis tournaments, namely: Roland Garros, BNP Paribas Paris-Bercy Masters and the Marseilles Open13.

Following an analysis of that part of recent writing on RBV which focuses on resource management, and having already factored in the BM concept, a longitudinal qualitative methodological approach will be put forward. From which will follow, consistent with a process of comparative replication (Yin, 1994), the analysis of the three sporting event cases and then a discussion about the contributions made by research.

2 Transition of the RBV Approach, from an Asset-Based to a Relational Perspective

One of the main difficulties presented by the RBV approach has been defining the key concepts represented by resources, competencies, assets, capacities and dynamic capabilities. The definitions used most often in RBV work present the notion of a resource as "stocks of available factors that are owned or controlled by the firm" (Amit & Schoemaker, 1993, p. 35). As for competencies, these are defined as the ability of the company to expand its resources by combining them while "key competencies" refers to the collective knowledge of the organization, especially its way of coordinating various production know-hows and integrating multiple layers of technology (Prahalad & Hamel, 1990). Finally, dynamic capabilities represent the company's aptitude to renew, increase and adapt its strategic skills (Eisenhardt & Martin, 2000).

If we take an interest in the development of RBV trends of thought, the initial work of Wernerfelt (1989) Barney (1991) Grant (1991) or even as Peteraf (1993) was mainly interested in the nature and properties which the firm's resources needed to have in order to be able to create and maintain a competitive advantage. According to this perspective, internal resources are analysed independently in terms of value creation potential, how hard it is for competitors to imitate or substitute them, or even their scarcity (Barney, 1991). Other theoretical ambitions aimed at identifying and operationalizing the isolating mechanisms (Rumelt, 1984) inherent to the firm's resources which are described as strategic (Peteraf, 1993) drove the work which followed. In the same vein, certain authors (Teece, 2007; Sirmon et al., 2007; Sirmon, 2009; Bingham & Eisenhardt, 2008) have recently underlined the importance of the interconnections and the complementarities between resources in enabling the creation of a network which is unique, gives value, is hard to imitate and is integrated into the organization of the firm. From then onwards, the perspective has focused on the importance of distinguishing competencies (Prahalad & Hamel, 1990) defined as the triggers for the deployment and combining of key resources (Holcomb, Holmes, & Connely, 2009), then on dynamic capabilities (Teece, Pisano, & Shuen, (1997)), those same distinguishing competencies to be renewed.

During the decade between 1990 and 2000, RBV trends of thought, based on competencies and dynamic capabilities, were elaborated as an alternative to Michael Porter's adaptive (or supposedly so) perspective arising from the industrial economy.

However, one of the inherent weaknesses of such approaches resides in their focus on the very nature of the resources and their failure to take into account the environmental players and factors which are close to the firm (Priem & Butler, 2001) and, as a consequence, in their all too frequently deterministic or mechanistic nature. The relative performance of a company would thus only depend on the characteristics of the owned and/or controlled assets and on the deliberate local activation of these characteristics.

At the end of the 1990s, the taking into account of environmental actors and factors through the concept of market-based assets (Srivastava, Shervani, & Fahey, (1998); Srivastava, Fahey, & Christensen, (2001)) brought about a return to a more balanced perspective where external issues once again had a place. These market or network assets are considered either according to their relational dimension (belief, reputation or customer relationship) or to their intellectual and cognitive dimension (knowledge of the competitive environment in terms of product offer and demand, consumer preferences). The strategic asset is then no longer a uniquely internal factor but feeds on and absorbs part of the environment close to the firm.

From an asset-based point of view, where assets can be controlled by invoking direct ownership rights (Miller & Shamsie, 1996), we then move towards a more relational, or interactionistic, point of view in which assets shared with components of the firm's environment (Rindova & Fombrun, 1999) may be controlled in the context of a dynamic strategic game. Successful external control (which is never guaranteed) can reduce a certain dependence on resources which are often co-constructed and shared with certain stakeholders (Coff, 1999; Pfeffer & Salancik, 1978) who are deemed free in respect of their own strategies and capable of making their demands known.

By integrating elements linked to the internal and external interrelations shared with the active components of the environment, the RBV approach brings about a relational renewal (Dyer & Singh, 1998) and connects with certain rereadings of the theory of the firm which attribute importance to direct stakeholders who are able to make their intentions known.

The taking into account of these intentions, which may affect the elaboration and roll-out of a firm's assets (Coff, 1999; Rindova & Fombrun, 1999) brings the RBV approach closer to the work of Pfeffer & Salancik (1978) on resource dependence and the external control of organizations.

2.1 The "Management View": Dynamic Arrangement and Control of Co-specialized Assets

Holcomb et al. (2009) underline the importance of managerial skills as a source of value creation through the creative capabilities of unique asset combinations. To that effect, Kor & Mahoney (2005) insist on the importance of manager action in terms of governance and deployment of resources which bring to mind the founding work of Penrose (1959, p. 5): "... the experience of management will affect the productive services that all its other resources are capable of rendering."

Henderson and Cockburn (1994, p. 77) also underline the importance of architectural integrative and combinatory competencies. For Helfat (1997) and Teece (2007), one of the priorities comes down to identifying complementariness (interfacing) and bringing about interactions (integration) between assets.

The work of Teece (2007) on the nature and the micro-foundations of dynamic capabilities is among the rare works which integrate the BM concept in a practical manner within a theoretical approach arising from RBV trends of thought. For Teece (2007, p. 1329, "the function of a BM is to articulate the value proposition." The terms "architecture" or "design" are mostly used to evoke the BM concept as a support for the orchestration of a firm's assets (Teece 2007, p. 1336). In the same perspective, one of the fundamental elements for formalizing a BM designed as a support representative of the way in which strategic assets should be arranged, concerns the management of asset co-specializations (Teece, 1986; Lippman & Rumelt, 2003; Teece, 2007). Co-specialized assets are defined as "a specific class of complementary assets where the value of an asset is a function of its use in conjunction with another specific asset" (Teece, 2007, p. 1338). In this way, the configuration of assets which are internal and external to the firm is carried out based on the identification and management of co-specializations in order to create a form of reciprocal "lock-in", a command unit which differentiates itself from the resource market but which benefits from the articulation of resources which are not owned.

At this stage, the BM concept appears as being able to help operationalize the various RBV approaches and especially the relational perspective (Dyer & Singh, 1998), which attributes importance to the management of strategic assets (Mahoney, 1995; Morgan, 2000; Sirmon et al., 2007, 2009; Holcomb et al., 2009) and to taking players in the firm's immediate environment into account (Coff, 1999; Rindova & Fombrun, 1999). In addition to Tapscott (2001, p. 5) who refers to the firm's architecture and particularly to the manner in which critical resources are deployed, other authors underline that one of the main contributions of a BM is its ability to show how the assets of a firm (Magretta, 2002, p. 91) are "configured" (Venkatraman & Henderson, 1998, p. 4; Hamel, 2002, p. 91) or "organized" while "making explicit the way in which an organization is linked to its external stakeholders" (Amit & Zott, 2001, p. 181). The works of Sirmon et al. (2008) as well as Holcomb et al. (2009) are the first to suggest an empirical development aimed at focusing on managerial action for arranging and deploying a firm's assets. Moreover, these recent works contain another special feature linked to the choice of the application ground which concerns professional sport: data for Major League Baseball (MLB) and the National Football League (NFL).

2.2 Business Model and Asset Combination

The term "Business Model" has been a part of the conceptual tool set of most consultancy firms for over 10 years, in spite of the vagueness surrounding its definition. However there is no consensus on the notion in the academic management community. We lack a clear definition which is accepted by everyone. Academic work has attempted to question the concept by putting it through the filter of major theoretical trends[1] (Chesbrough & Rosembloom, 2002) in strategic management. But these connections and a posteriori theory transplants create problems. The notion of a BM, which was defined too widely and too quickly, encapsulates assumptions and a model of organized human action which still deserve to be questioned. Starting from definitions which appear in writing on subject, we have tried to build the smallest possible common denominator of the notion and are proposing in the end to define a BM as "the choices a company makes to generate revenues". Even though this first definition has the merit of being simple it remains extremely broad and asks several questions as to the types of choices accessible by a particular company within a competitive industry and context. This is the main criticism of Porter (2001) who considers that the BM represents one more attempt to explain in a summary and quick way how the company generates profits, sacrificing the complexity of reality and the contingency of choices. By putting forward its lack of theoretical foundations, Porter invites the researcher to avail himself of the methodological caution which the firms in charge of its promotion appear to have lacked.)

The aim is to submit to analysis a clear representation of the architecture of a firm and to try and show how all its critical resources are deployed and controlled (Tapscott, 2001; Kor & Mahoney, 2005; Sirmon et al., 2009; Holcomb et al., 2009). The question to ask is how to arrive at this kind of modelling and how to build such a BM by adopting the normative stance of the architect. In this perspective, Hamel (2002, pp. 69–118) approaches the BM in a very operational way by dismantling the integrative construct which it represents and by carrying out a rereading in terms of a tool set close to the engineering of organizations.

At first, Hamel (2002) suggests rebuilding the BM concept from four major organizational components: the client interface, the strategic core, strategic resources and the value network. The interest of Hamel's approach, taken up by Shafer, Smith,

[1] Convention theories, Resource-Based View approach, stakeholder theory.

and Linder (2005) lies in its highlighting the articulations which may exist between these components. Three types of organizational coordination mechanisms, which correspond to as many architectural decisions linking the BM components together, are thus put forward.

These three mechanisms or decisions (configuration of activities, client benefit, firm borders) are considered as the foundations and the guarantors of a BM' s rationale. The configuration of the activities refers to the way in which the competencies, the assets and processes are "combined" and "interlinked" in order to implement a particular strategy. The aim is to put forward articulations between competencies, assets and processes and especially the way in which these articulations are managed. Client benefits are defined as the elementary needs to satisfy in a target market. The borders of a firm represent the decision-making scope of the company within a micro-environment where suppliers, partners or members of a coalition intervene in terms of the production and value chain. This scope of decision corresponds in practice to the scope of control which the firm enjoys in terms of the resources shared with all of the various stakeholders.

3 Method: Three Clinical Studies

One of the difficulties inherent to the RBV approaches is methodological (Rouse & Daellenbach, 1999, 2002). But the writing regarding this difficulty (Godfrey & Hill, 1995; Rouse & Daellenbach, 1999, 2002) can help us to formulate a longitudinal operating mode[2] on the basis of studies of clinical cases (Eisenhardt, 1989, p. 533) selected in this case on a theoretical rather than statistical basis.

In focusing on the management and control of the internal and external resources of the firm and on their integration within a coherent BM we are looking to observe and detail an internal process which is supposedly rational. Our objective is to investigate the ground favourable to the observation of a deployment of the BM linked to problems of organization (Sirmon et al., 2009; Holcomb et al., 2009) and to control of the assets of a company (Hamel, 2002). One of the facets of a BM is then considered as a fundamental support to the creation and maintenance of a coherent group of assets articulated according to their co-specializations (Teece, 2007) within a certain scope of control (Freiling, 2008).

The following table summarizes the choice of cases selected according to the theoretical criteria identified in RBV writing (Table 1).

Our approach will follow a sequential collection and analysis process adapted from the works of Morgan (2000) and Sirmon et al. (2007) relating to the construction and deployment of a portfolio of resources. At first we will marshal the principles of the "naturalistic" perspective (Lincoln & Guba, 1985), coupled with analysis techniques arising from the approach known as the "grounded theory" (Strauss & Corbin, 1998). In a second phase, a more participative approach to action research

[2] The investigation began in 2001 and continued until 2008 following the approach detailed above.

Table 1 Selection of cases based on theoretical criteria

Choice criteria of cases	Open 13	BNP Paribas Masters	Roland Garros
1/Belonging to the same industry (Rouse & Daellenbach, 1999)	Tennis industry and events		
2/Presence of differences in basic financial indicators (Reed & DeFilippi, 1990; Rouse & Daellenbach, 1999)	Budget: €4 million	Budget: €10 million	Budget: €125 million
3/Endowment with resources of a similar nature (Godfrey & Hill, 1995)	Sporting-partnerships (sponsoring—Public relations)—venue (stadium)—		
4/Resources interacting with the components of the micro-environment (Rindova & Fombrun, 1999; Coff, 1999)	Suppliers of sporting events (athletes)—technical suppliers—partners—federation—local authorities		
5/comparable BM at the architectural component level (Hamel, 2002; Teece, 2007)	Comparable BM with a different organization of the components		

(Whyte, 1991) whose objective is the cogeneration of a learning curve (Greenwood & Levin, 1998) between actors on the ground and researchers is used.

The following table retraces every step of the methodological protocol implemented while underlining the observation methods and the analysis techniques.

Following the presentation of the methodology implemented, each case will be presented according to the sequential approach: Identification—prioritization—deployment—organization—control—and rationale of the BM implemented.

3.1 The Open13. A Relational Control of Resources Shared with the Main Stakeholders

The Open13 is a professional tennis tournament in the "ATP International Series" category which takes place in Marseilles every year in mid-February. The event first took place in 1993. It belonged to the IMG McCormack group[3] until 1999 when its director, Jean-François Caujolle (JFC) was given the opportunity to acquire the event and to continue to organize it on his own by founding a family company (Pampelonne Organization).

The Open13 is the last French international tennis tournament to belong to a small to medium-sized family company. JFC, its director, is also the chief

[3] IMG is one of the largest sports marketing agencies in the world; its activities are associated with the management of TV rights, image rights for major sports figures (Tiger Woods, Michael Schumacher, Roger Federer, ...), international celebrities (show business, cinema) and the management of sporting events.

executive of one of the main French sporting and cultural Public Relations agencies (Pampelonne). The history of the tournament is linked to the history of JFC, a former high-level tennis player in the 1970s, who is originally from Marseilles. This self-taught entrepreneur has built up this tournament based on two principles: a principle of trust by surrounding himself with family members and friends, and a well-thought out partnership principle through the support of the event's main partner, the Conseil Général des Bouches du Rhône. A platform of major private groups who are official partners (BNP Paribas, Sodexo and Peugeot) and national and local companies with a highly diverse range of businesses round out this relationship structure.

Analysis of the archives and the meetings held in the context of research has highlighted four broad types of resources. Each one of these is characterized by two key elements: The contribution of one or more stakeholders to its construction and its link to a profit center. Table 2 describes all these elements.[4]

Two categories of resources were mentioned most often during the interviews conducted:

- The reputation of the event, which is the very essence of the tournament as it attracts both spectators and partners.
- The social capital of the tournament whose representative forms (relationships, personal networks, public relations, personal acquaintances) are mentioned most often on a quantitative basis.[5] The partnership and physical resources categories were mentioned less frequently by the interviewees but were still very often associated with the social capital of the event and its reputation as a sporting event which is suitable for Public Relations.

Once the representativeness of the resource categories has been put forward, the question of the deployment of those same resources (Grant, 1991) via certain distinctive skills (Pralahad & Hamel, 1990) represented the last type of analysis in the interviews conducted.[6]

Two main types of distinctive competencies have been emphasized, both of which were originally directly linked to the tournament manager (JFC):

- The ability to offer an attractive array of players every year which required risk-taking in terms of offering financial guarantees to promising young players (examples: Roger Federer in 1999, Raphael Nadal in 2005, Novak Djokovic in 2006, Andy Murray in 2007 as well as Jo-Wilfried Tsonga in 2009), by attracting

[4] The profit centres of a sporting event mainly involve revenues generated by partnership contracts (sponsorship, public relations, public subsidies), media rights, ticket sales as well the sale of tie-in products and merchandising (Maltese, 2008).

[5] Lexical forms which were present most often during the interviews.

[6] This type of analysis refers to the selective coding techniques described by Strauss & Corbin (1998) aimed at coding certain elements of the respondent's speech in relations to themes identified in writings on dynamic competencies and capacities.

Table 2 Categorization of the Open13 resources

Resource category	Description	Links to profit centers	Stakeholder contribution
Partnership	Public (subsidies) and private (sponsorship) sponsor contracts including service exchanges (catering, security, transport...)	Main sources of tournament revenue: main official partners and various sponsor services (printed material, media, visibility on the tennis court, private areas, ticket purchases...)	Major contribution by main partner CG13 in terms of financing (1/4 of the budget) and by official partners who often act as suppliers (Peugeot: transport, Sodexo: catering, Onet: security, cleaning...) French Tennis Federation through the Provence League: access to the regional tennis-playing community.
Social capital	Social network of the directors involved—Public Relations Services, which creates and event social network (decision-makers of major partnership groups—mayors, directors of the event and the French Tennis Federation)	Marketing of Public Relations transactions with all types of companies: – Top-of-the-range ticketing – Catering – Private areas – Access to transport, parking, hotel facilities...	Contribution of decision-makers in partner companies who often act as "middle-men" or ambassadors for the tournament's directors. Contribution of certain directors of the French Tennis Federation to the negotiation of public and private partnership contracts.
Reputation	Reputation of the event (local profile, history), of the players (first and foremost of the professional players via their managers), of the tournament manager (former high-level sportsman, known in local sporting and economic circles)	No direct link with a significant profit center other than the rights linked to communication platforms (especially media), which remain marginal for this type of event (less than 3% of the budget).	Strong contribution of professional players (in this case, the players are attracted into taking part by financial guarantee contracts) who are the event's main attraction. Contribution of main and official partners in terms of the potential attraction of the venue for other companies.
Physical dimensions	Stadium (Palais des Sports in Marseille—Venue (completely redesigned by the organizer for the duration of the event). Local potential in terms of spectators	Ticketing (all price ranges) Hospitality (parking, mass catering).	Contribution by the town of Marseilles (owner of the Palais des Sports). Local ticket distribution networks.

the maximum number of French players or high-charisma players. The player line-up is the tournament's advertising and requires the continuous involvement and investment (during and outside the competition) of the event manager and his team in terms of their relationship with the players' agents, the monitoring and anticipation of performances and especially of a relevant analysis of the personal calendar of each sportsman or woman in terms of their tournament choices each year.
- JFC's social capital in and outside sport allows him access to a quantity of personal information on the executives of large groups and local authorities ready to invest at the advertising level. These relational skills (Blyler & Coff, 2003) guarantee his colleagues privileged access to the same decision-makers, who are partners or future partners, thus enabling the roll-out of customized offers which are the reason for the loyalty of most clients who sponsor the public or private event.

In the end, the organization of the tournament deploys its resource base (Hamel, 2002) with all the tournament assets through the following capabilities:

- Renewal of certain contacts and contracts based on the personal information which JFC manages to obtain in his professional network, both from a sport and partnership standpoint.
- Commercial use of the social and relationship capital associated with JFC and the tournament in order to create and strengthen close links with partners (private and public sponsors) and suppliers (especially the players as well as the managers of the venue (Palais des Sports) which belongs to the town of Marseilles).

Being part of the organization of the event has allowed researchers to observe directly the way the different assets detailed above are arranged within a perspective of co-specialization of the tournament's resources. One of the central characteristics of this event is directly linked to the managers' choice to concentrate both their human and technical investments on activities associated with their resource base. In this way, prospecting for new partners and keeping existing partners loyal is carried out through the putting in place of an approach based on the interpersonal relationships of the members of the organization. Contracts where the content refers to quantifiable returns in terms of advertising or marketing are rare. It should be noted that around 80% of tickets are sold to companies, who invite their clients or employees. Here again, the social network of the tournament team enables value to be extracted from this profit centre. Finally, as far as the Palais des Sports and the venue is concerned, the relationships which the tournament manager and his technical director enjoy with the employees of the town of Marseilles in charge of the event are the reason why Public Relations areas (partners' village), built exclusively for the tournament, were created. This village and these spaces have become supporting platforms which are strongly identified with the Open13, creating a certain uniqueness[7] for top-of-the range partnership offers.

The resource base (Hamel, 2002) on which the other strategic resources are articulated includes the reputation of the event and the social capital of the directors. However, both these asset types are directly dependent on the contribution of certain stakeholders. This is the case of a reputation which is highly dependent on the presence or absence of the best players on the circuit.

The analysis of the BM implemented by the event managers enables us to understand, based on the arrangement of key assets, how an organization which is affected by particularly unstable local sporting and socio-economic environments can control certain intangible resources which have been co-constructed with certain determining stakeholders.

Figure 1 is a stylized illustration of the BM implemented by the Open13 management. We will use it as a tool for analyzing the intrinsic difficulties of controlling and deploying the assets.

This illustration enables us to highlight the choices that the event organization makes to generate revenues which are essentially linked to partnerships. The social

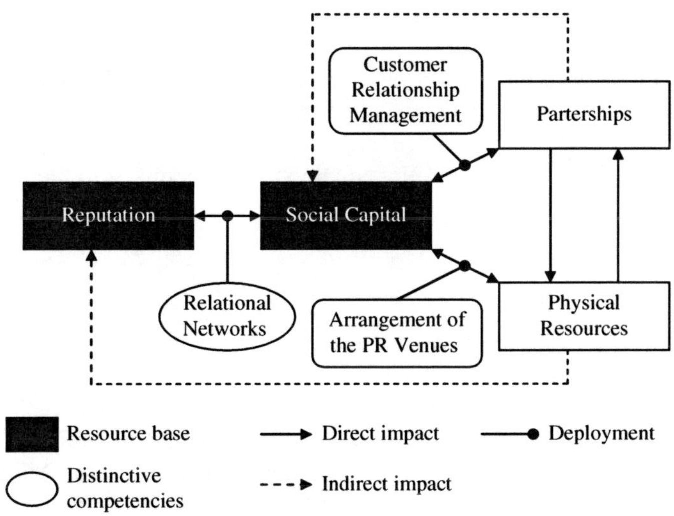

Fig. 1 The "Open13" business model

[7] The creation of this Public Relations Village dedicated to the tournament's main partners has become a priority for the organizers and the decision-makers in the client companies. The differentiating, even unique characteristic of this space concerns the choice of its temporary layout, of the themes and decorations put in place from a technical standpoint which enable the creation of a convivial physical support platform for a week.

capital of the directors plays a pivotal role in the that the relational networks of JFC and his team are the reason for:

1. the presence of the best players which guarantees the sporting reputation of the tournament,
2. the closeness of the partners symbolized by a particular effort to maintain relationships throughout the year with the main decision-makers at the event partners,
3. the creation and the development of a specific venue (VIP Village) enabling the commercialization of the Public Relations areas which play a physical support role for the social networks developed around the tournament.

In this way, by returning to the work of Hamel (2002) on the elements which form a BM, the construction rationale of the BM of this event is as follows:

1. the benefits to the main clients who generate revenues relate to the providing of public relations[8] and access to one of the main regional social networks,
2. the configuration of strategic resources hinges on the combination of sporting reputation (Parent & Foreman, 2007), fames (Hayward, Rindova, & Pollock, 2004) and the social capital of the directors (Blyler & Coff, 2003)
3. the scope of control of the strategic resource base corresponds to the extent of the directors' social network both for partner relationships and for suppliers and coalitions with other sporting organizations.[9] The evolutionist perspective (Durand, 2006) of such an organization therefore depends on the capacity of its managers to manage and make commercial use of their relationship capital (Granovetter, 1973) without exceeding certain size limits linked to the management of strong and weak ties.

3.2 The BNP Paribas Masters. A Control of Resources, Which Are Not Shared and of Sporting Dependency

The BNP Paribas Masters (BNPPM) will celebrate being staged for 23rd time at the end of October—beginning of November 2008. This tournament is organized by the French Tennis Federation (FFT) and ranks among the nine Masters Series on the ATP circuit. These types of tournaments have a number of specific characteristics, including having common partners and enjoying the participation of the

[8] Comercialization of the Public Relations services represents over a third of the event's revenues.
[9] In 2008, at the end of December, The Open13 group Canal Plus events (which owns the Lyon ATP tournament) and the FFT (owner of the BNP Paribas Paris-Bercy Masters) created a tournament which brings together the best French players (Masters France) and which takes place in Toulouse from the 18th to 21st of December, enabling common partnerships (the ACCOR group, for example) to be generated.

best players in the world[10] without the organization giving out guarantees (financial incentives), unlike tournaments in the Open13 category. This event takes place annually at the end of October-beginning of November over a week in the Paris Bercy Palais Omnisport (POPB).

The BNPPM has had less exposure in its last few seasons because the best world players were not taking part at the end of the season. In addition, the new governing body of the Association of Tennis Professionals (ATP) introduced a certain number of reforms from 2006 onwards, concerning the tournament calendar in order to reduce the number of competitions and re-energize the circuit. The Masters Series ranking will thus disappear from 2009 onwards in favor of other category types.The BNPMM, given its weaker results in terms of attendance (less than 100,000 spectators) was already in a delicate position in 2006 compared with other tournaments as far as continuing to benefit from a ranking which would guarantee the presence of the best players is concerned.

In this perspective, the FFT decided to appoint a new tournament director and set him the task of re-energizing the event and facing up to the threat of relegation hanging over the placeCityParis competition in particular. The FFT chose to appoint JFC (the manager of the Open13) from 2007 onwards.

One of his first initiatives was to obtain permission to start the tournament on a Sunday and to organize a private concert[11] to launch the event. The concert was financed by the main partner (BNP Paribas) which invited 8,000 young people to this event called "Sunday Start[12]"

The analysis of the interviews conducted with all of the players, both organizers or decision-makers, among the stakeholders of the event followed the same approach as for the Open13, and enabled a similar categorization of the resources of the tournament to emerge, which is summarized in Table 3.

The resource base on which the event relies to differentiate itself has changed significantly over the past 2 years. Its reputation, especially in the world of sport, has always been its major asset in terms of attracting spectators and companies (Fombrun & Van Riel, 2004) but the support of the main and official partners has been the key resource, especially in terms of financing. The origin of these partnerships was both the result of the sponsorship policy common to all nine Masters Series tournaments combined with partnerships attracted mainly by the other major event organized by the FFT, Roland Garros. For the last 3 years, the spectator-fan has been at the heart of the BNPPM's event strategy with, in particular, an optimal use of the venue of the POPB in terms of entertainment, interludes and services which are peripheral to the main sporting event.

[10] The most highly ranked players are obliged to register for this tournament category and can only refuse to participate by invoking health reasons.

[11] Concerts with famous DJs: David Guetta in 2007 and Martin Solveig in 2008.

[12] The "Sunday Start" enabled the organizers to involve the main partner and mainly to communicate the repositioning of the event's identity, combing a high level of sport and offering entertainments (Entertainment) associated with the tennis event.

Table 3 Instrumentation of research

Stages	Observation methods	Analysis techniques
Phase 1: Immersion—naturalistic approach—theoretical grounding		
1. Identification and categorization of assets	Collection of secondary data: archives—press and marketing files—internal documents—budgets—internal and external media	Open coding: categorization of the key factors for success by completing manual coding (Nvivo software package) starting from redundant themes with an Alceste lexicometric analysis (Reinert, 1990) enabling the quantification of joint lexical methods
2. Prioritising of resources	Collection of primary data: open meetings (Holtsein & Gulbrium, 1995) conducted with decision-makers (28 in total lasting between 1.00 and 1.30, fully transcribed between 2001 and 2004) for the three tournaments as well as with decision-makers for all the stakeholders involved (media-private and public partners-federation - suppliers)	Axial Coding: coding of the ties which exist between different resource categories (Nvivo manual coding) combined with the quantification of key words representative of every resource category identified at 1 (Alceste lexicometric analysis)
3. Deployment of resources		Selective coding: manual coding linked to the types of competencies identified in RBV literature—Dynamic capabilities and those put forward by the respondents. The objective of this coding is to highlight the ability of events to explore and make use of their internal and external resources
Phase 2: Action research		
4. Organization and co-specialization of the assets	Integration within the organization of events (from 2004 to 2005 onwards) as a development consultant by collaborating directly with the tournament directors. Our activities have been dedicated mainly to the strategic repositioning of events, to changes to Business Models and the restructuring of the organizational teams as well as to technological innovations	Participation in the events management meetings and in the carrying out of assignments dedicated to the reconfiguration of the event asset portfolio and the reorganization of the teams since 2005
5. Scope of control and resource dependence		Participation in partnership negotiations and the formalization of partnership contracts (local authorities, private sponsors, suppliers, media, athletes) since 2005
6. Global process and implementation through the BM deployed		Integration within the committee of experts dedicated to the repositioning of events and the changes to Business Models since 2006

This policy is also directly linked to the marketing of tickets for every kind of consumer including multiple offers linked to tools for the management of the direct marketing client relationship. The BNPPM resource base thus includes the tournament's reputation, which plays the role of "attractor" or "trigger" combined with the physical resources representative of the use of the POPB venue supported by the policy of optimizing the ticket offers (Table 4).

In this perspective, changes are seen in the partnership strategy in terms of commercial arguments supplied by the organizers focused on the content of the tournament both in terms of the event on offer and of the quality and quantity of spectators present.

As far as the social capital of the event is concerned, that comes last, given the degree of externalization of the Public Relations activity.

In terms of the deployment of strategic resources, three distinct types of aptitude are highlighted:

- Managing the sports side by creating a dedicated unit led by JFC in order to monitor the players' needs throughout the year both in terms of the surface on which the game is played, the type of balls used and the quality of the sportsmen and women's welcome and stay in Paris, as well as that of their family and their team. This ability enables the deployment of the tournament's sporting reputation.
- A more "experience-based" production of the event combining the staging of a high level of sport with peripheral and entertaining shows. The sporting event is thus considered from an Entertainment angle which aims to give the tournament a new identity supported by the venue's potential and the contribution from certain specialist suppliers within the POPB. This aptitude enables the deployment of physical resources and also of the tournament's global reputation.
- Ticket sales aimed at widening targets in both individual and company terms and proposing customizable offers. This ability is directly linked to the commercial use of physical resources and the local potential in terms of the consumption of a sports event which is both of a very high level and entertaining.

The co-specializations between assets are articulated around the physical support and the quality of the sporting line-up. Indeed, as we mentioned earlier, the sporting venue has an impact on whether the best players attend. We note that, in the context of the extension of the Roland Garros stadium, the FFT is allowing for the creation of a covered tennis court which may potentially serve as a new modern venue for the BNPPM. Partners, in addition to the ranking and the quality of the sports line-up, are equally interested by the capacity of the organizers to attract more and more spectators from extremely diverse socio-economic backgrounds. The main partner, BNP Paribas, is also financing the opening of the tournament and putting on a concert which will allow advertising to new targets (young people and students in the Paris region). The diversity of the spectators present in the POPB tribunes, combined with the investments made to reinforce and broaden the high-level sports event around many entertaining events highlights the pivot role of physical resources, with a perspective of total control via the construction by the FFT of the new enclosure in the Roland Garros stadium.

Table 4 Categorization of the BNP Paribas Masters resources

Resource category	Description	Links to profit centers	Stakeholder contribution
Partnership	Public (subsidies) and private (sponsorship) sponsor contracts including partners with a presence in all nine Masters Series tournaments	Main sources of tournament revenue: main official partners and various sponsor services (printed material, media, visibility on the tennis court, private areas, ticket purchases...)	Major contribution by main partner BNP Paribas in terms of financing and by official partners who often act as suppliers (Mercedes-Benz: transport, Sogeres: catering, Canal Plus: TV production-broadcasting...)
Social capital	Tournament managers social network—Internal FFT service—Public Relations Services. Note that for this event, the FFT calls upon 11 official Public Relations agencies to commercialize these services	Commercialization of Public Relations transactions with all types of companies: – Top-of-the-range ticketing – Catering – Private areas – Access to transport, parking, hotel facilities...	Strong contribution by the official suppliers represented by the 11 agencies (external to the FFT) approved to commercialize the Public Relations Services
Reputation	Reputation of the event (local national, international profile, history), of the players (first and foremost of the professional players), then of the tournament organizer, the FFT and its head, JFC	No direct link with a significant profit center other than the rights linked to advertising platforms (especially television media)	Strong contribution by the professional players who are the main attraction of the event. Contribution by the media partner (Canal Plus) in terms of TV exposure during the week in which the event takes place
Physical dimension	Stadium (Palais Omnisport Paris Bercy—POPB) whose venue is adapted to the "Entertainment"-type production desired by the organizers	Ticketing (all price ranges) Hospitality (parking, mass catering)	Contribution by the POPB employees and suppliers specialising in the technical production of events (GL Events, TAV...) and of the providers of public entertainment (Martin Solveig (DJ set)), Fred Viktor (creating the atmosphere for the event), Exyzt (video animations) and the "Bronx Drums" (urban percussion) for the finals

In contrast to the Open13, the scope of control of the strategic resources of the BNPPM is more extended, given the fact that the rank of the tournament must be maintained in order to make the attendance of the best players on the circuit more likely. The commercial use and the development of the sporting enclosure in terms of ticket marketing and event production must be kept under control for this reason.

The contribution of professional tennis players is essential to boost the reputation of the event through its line-up. The use of the potential of the POPB enclosure for the production of the sports and non-sports event is also a priority. In this case, the aim of that control is to limit the dependence on sport and offer a complete show which mixes very high level tennis competition and peripheral entertainments (concerts, mass audience entertainment, setting the scene for the matches, audience participation...) However, the heart of the offer remains the quality of the sporting line-up, which acts as the main attraction and reputation trigger. The organizer thus remains dependent on the choice of sportsmen and women and his independence remains marginal. In 2008, the withdrawals through injury of the two main players (Rafael Nadal and Roger Federer) for the quarter finals of the tournament were an illustration of this phenomenon.

The BM (Fig. 2) put in place for this event by the FFT illustrates the organizers determination to exercise control by placing physical resources at the center of the choices aimed at generating revenues. Reputation remains the main attraction point of the tournament and depends on the presence of the best players, but the potential for commercial use of the enclosure and ticket marketing remains under the control of the actions taken by the FFT.

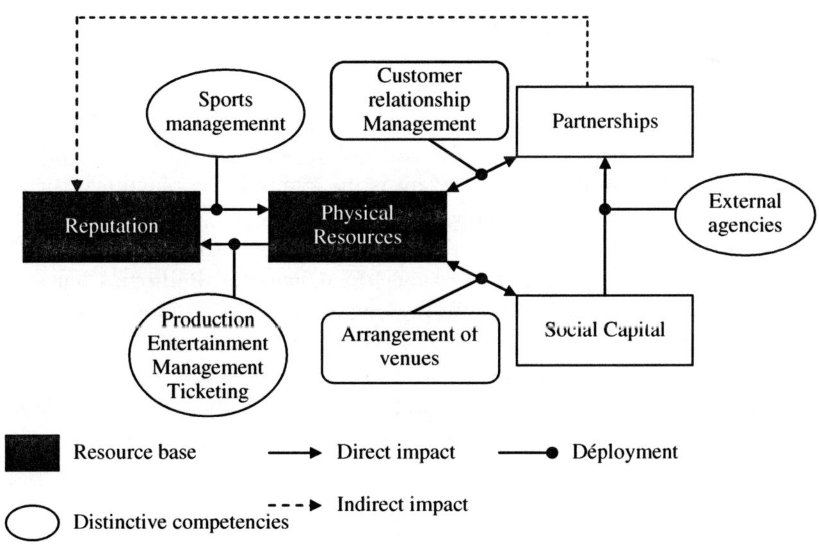

Fig. 2 "BNP Paribas Masters" business model

Client benefits are directly linked to the production quality and the sports event on offer, and also to the level of sponsorship by private companies and local authorities, to direct exposure (target of over 100,000 spectators) and the diversity of the Ile de France spectators attending.[13] The arrangement of resources is thus articulated around the combination of reputation—physical resources. For the 2009 tournament, two official partners (Mercedes Benz and Sagem) will not be renewing their contracts, the organizers' new partner prospection strategy will be directly linked to the arguments set out above. Through the commercialization of the Public Relations services, the social capital side remains associated with offers for tender made by the FFT to appoint specialist agencies. In contrast to the Open13, this activity remains secondary although it generates revenues.

The weight of the main partner remains considerable; its involvement in the financing of the "Sunday Start" is a perfect illustration of this. These final elements show that the BM can still evolve in order to make the BNPP completely independent, particularly in relation to the other major event organized by the FFT (Roland Garros) while making use of certain coalition possibilities in terms of a combined ticket offer at these two federal events.

3.3 Roland Garros Control Through the Brand, Creation, and Maintaining the Dependence of the Main Stakeholders Involved

Roland Garros (RG) is one of the two most important annual sporting events organized in France, along with the Tour de France. Just like the BNPPM, the FFT organizes this event at the Roland Garros stadium at the end of May-beginning of June. In addition to its "Grand Slam" ranking, which it shares with Wimbledon, the US Open and the Australian Open, the tournament, which is more than 100 years old, keeps its legend alive through its stadium. Known at the beginning for the exploits of the French players nicknamed *the Musketeers* in the 1920s during the Davis Cup (team matches between nations), the event is the symbol of a playing surface: red clay, which makes it a worldwide benchmark.

RG, the jewel in the FFT crown, is the Federation's main source of revenue. From the 1970s onwards, on the initiative of the FFT Chairman, Philippe Chatrier, RG has become more professional by offering communication products (sponsorship) to companies and opening ticket sales to every type of client. Today the unique nature of the event, linked with its rank and its legend, requires the extension of the stadium given the level of surplus demand. The opportunity to stage the Olympic Games in Paris in 2012 had enabled a initial solution for extending and modernizing

[13] In this case, one of the main commercial arguments of the event organizers is associated with the quantitative and qualitative potential of the specators present in the week of the tournament. This argument refers to the capability of the commercial managers to identify the people present who could interest a company within the context of its sponsorship operations.

the stadium. The failure of this bid pushed back the project, which remains a priority for the development of the event in line with the other Grand Slam tournaments which have all extended and modernized their enclosures.[14]

The consequences of this extension will be the construction of a new court with seating for over 15,000 people with a retractable roof, which will both allow part of the event to be ensured in case of rain and also, as we have mentioned, the ability to organize the BNPPM in the dedicated venues.

In terms of organization, the teams are the same as those of the BNPM except as far as the management of the tournament is concerned where the former director of the FFT, Jean-Claude Blanc, played the role of director of the event during his period of activity at the FFT, from 2001 to 2006. His replacement, Jean-Francois Vilotte, has kept the organizational framework of the RG tournament without a director being officially named.

One of the main strategic choices, besides the need for extending the stadium, was made under the management of JC Blanc and concerned the number of main partners and the management of the Roland Garros brand. JC Blanc's aim was to make RG independent in relation to its partners (companies and media) by trying to balance the resource portfolio and the associated profit centers. The event brand strategy implemented was thus conducted along two lines (Table 5):

- Make RG more exclusive in terms of marketing to companies;
- Open RG up to the greatest number of spectators and television viewers and then deploy the commercial brand in terms of merchandising and tie-in products by co-branding a certain number of products.

The analysis of the meetings conducted for the RG case study has enabled us to categorize the RG resources in the same way as previous events.

The resource base, as in the two previous cases, relies on the power of attraction (Fombrun & Van Riel, 2004) of the event's reputation. Contrary to the other tournaments, however, RG's "Grand Slam" position means that in this configuration it is not the players which make the event but the event which makes the player. In other words, in this case the players depend on the tournament in terms of boosting the value of their career and their ranking. RG is an obligatory stage for any professional player, in the same way as the three other tournaments in the Grand Slam. Just like the BNPPM, physical resources, which include the RG stadium and ticket marketing at the national and international level, are integrated in this resource base. Partnerships come next, according to a brand partnership rationale as well as the social capital symbolized by the Public Relations services.

[14] Since October 11, 2008, following the decision of the Paris Council on September 29 last year, the FFT has decided to launch an international design competition for the extension of the Roland Garros stadium in the municipal Georges Hébert stadium at the Porte d'Auteuil.

Table 5 Categorization of the Roland Garros resources

Resource category	Description	Links to profit centers	Stakeholder contribution
Partnership	Public (subsidies) and private (sponsorship) sponsor contracts including service exchanges (catering, security, transport...)	One of the four major sources of revenue of the tournament: main official partners and various sponsor services (printed material, media, visibility on the tennis court, private areas, ticket purchases...)	Major contribution by main partner BNP Paribas in terms of financing and by official partners who often act as suppliers (Peugeot: transport, Orange and France Television: TV production-broadcasting...)
Social capital	Tournament managers social network—Internal FFT service—Public Relations Services. Note that for this event, the FFT calls upon 11 official Public Relations agencies to commercialize these services	Commercialization of Public Relations transactions with all types of companies: – Top-of-the-range ticketing – Catering – Private areas – Access to transport, parking, hotel facilities...	Strong contribution by the official suppliers represented by the 11 agencies (external to the FFT) approved to commercialize the Public Relations Services
Reputation	Reputation of the event (international notoriety, history and legend), and of the participants (with the professional players in first place). Grand slam toumament category (along with Wimbledon—the US Open—the Australian Open) World red clay championship	National and international TV rights Commercial and tie-in products brand (co-branded with Adidas in particular and distributed internationally)	Contribution of media partners and international broadcasters. Contribution of supplier partners and particularly Adidas in terms of design and equipment branding of the Adidas-Roland Garros equipment range
Physical dimension	Roland Garros stadium and "operation Roland Garros in town" whereby zones for retransmitting the matches are set up in Paris and in other towns.	Ticketing (all price ranges) Hospitality (parking, mass catering)	Contribution of local authorities to the extension of the site and putting in place Roland Garros operations around the town. The Roland Garros stadium is directly licensed by the FFT

In terms of deploying the strategic resources previously identified and according to the same logic as the "mega" events, two types of ability appear distinctive:

- The management of the stadium, its modernization and its commercial use as a direct link to the ticket marketing strategy aimed at maximising this source of revenue and making use of the information it holds about its licensees—customers—fans. According to this rationale, the aim is to supply other markets: attracting and retaining sponsors, merchandising, tie-in products, on-site consumption, media, multimedia, telecommunications and video games.
- The management of the Roland Garros brand both in terms of the Business to Business market (partnerships and Public Relations) and the commercialization of Roland Garros products nationally and internationally.

In this case, co-specializations are integrated into the Roland Garros brand which is also the name of the stadium. Indeed, the management of the reputation of an extremely expressive organization (Schultz, Hatch, & Holten Larsen, 2000) allows the commercial use of a brand whose symbol is the RG stadium. The stadium is thus interchangeable with the event, as if it were named after a town. The attraction and the capacity of the organizers in terms of managing the client relationship (ticketing strategy) both at the spectator and company level, combined with the choice of maximum media exposure, reinforce the potential for attracting and maintaining private and public partners. Finally, this same brand attracts companies which consume public relations services.

The scope of control of the key assets depends essentially on the capacity of the FFT to develop and make use of its brand and its stadium. In this case, the reputation-physical resources combination is not constrained by the non-controllable contribution of certain stakeholder suppliers, since the high level players are dependent on this type of tournament and the stadium is the property of the FFT. The potential for the commercial use of the resource base thus has very few limits. Only the uncertainties linked to the business of sport can slow RG's momentum. The doping related controversy which currently hangs over the Tour de France, for example; the threat of corruption (Italian professional football, basketball, American NBA); or again the potential revolution brought about the liberalization of online betting, to which the FFT is extremely attentive, are major challenges which the managers of the FFT will have to face to protect the reputation and sustainable use of the RG stadium.

Figure 3 shows the current RG business model in terms of the different choices made to generate revenues from strategic resources and the distinctive abilities associated with the event. The essential difference compared to the previous BM is the low level of impacts which are indirect or over which the organizers exercise little control. In other words, the contributions of certain stakeholders (particularly the media and partners) are the result of choices made by those organizing the tournament. Indeed, the client benefits are first and foremost linked to the exceptional nature of RG. The configuration in management terms of all of the resources of the event hinges on its reputation and the RG brand. Finally, the limits of RG are

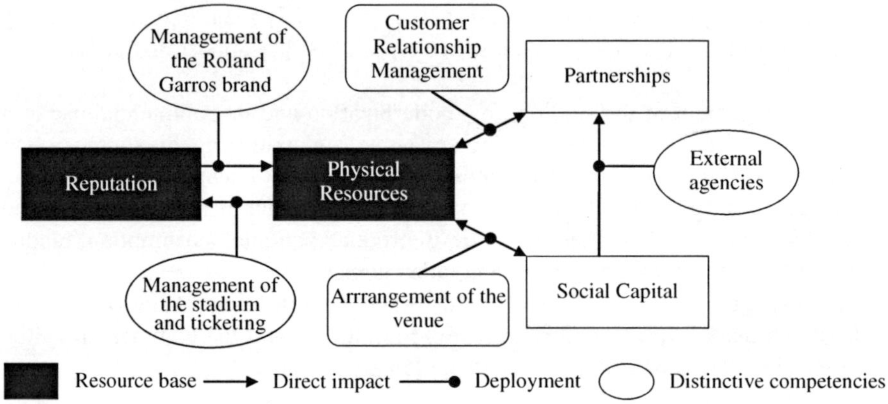

Fig. 3 "Roland Garros" business model

associated with the capability to internationalize the reputation and the brand of the event, both in terms of partnerships and media rights and of ticket marketing, merchandising and tie-in products.

4 What the Clinical Studies Teach Us. Differentiated Combination and Control of an Identical Resource Pool

The studies of clinical cases of tennis sporting events conducted here enable the highlighting of two types of contributions associated both with the BM business model and with the RBV research trend centered on the management of the firm's resources (Mahoney, 1995; Sirmon et al., 2007, 2009; Holcomb et al., 2009).

One of the results of the clinical study of French tennis tournaments involves the choices of configuration of each BM based on a globally identical base of strategic resources (Hamel, 2002). The types of resources[15] are the same, only the allocations change depending on the importance of each event (Holcomb et al., 2009). Another common characteristic, inherent to the types of companies studied and which produces sporting events is linked to the importance of reputed resources as the main attraction points (Fombrun & Van Riel, 2004) of the other resources. The reputation of a sporting, artistic or cultural event refers to the quality of the line-up drawn up by the organizers of the live event. For the Open13, this depends strongly on the financial (contracts of guarantee for the attendance of the best players) and relational capacity of the organizer with the sportsmen and women. For the BNPPM, the sporting relationships of the organizers also play an important part but it is mainly maintaining the rank of the event (Masters Series) which guarantees the quality of the sporting line-up. As far as RG is concerned, its associated Grand Slam status

[15] Partnerships, reputation, social capital and physical resources

and its history make sports figures dependent on the event in the sense that their own reputation will be built depending on their performance in such a tournament.

However, at each event, the strategic choices of configuration of the BM starting from a pivotal resource directly linked to the event reputation are different.

In the case of the Open13, the main source of revenue resides in the Public Relationships partnerships and transactions. The tournament director has made a choice to invest in its social capital and in that of his team to build his BM. However, control of the scope of the resource base (reputation–social capital) is highly dependent on the capacity of the organizers to attract the best players[16] and to maintain and develop their interpersonal relationship networks.

The directors of the BNPPM have not made the same event-driven BM choice. Although the main sources of revenue are equally linked to partnership contracts, physical resources[17] (stadium and ticket policy) are positioned as the cornerstone of the tournament's BM (Peteraf, 1993). In this case, control is based on the capacity of the organizers to attract, retain the loyalty of and diversify the audience while making commercial use of the potential of the sporting enclosure.[18] The potential for control, though wider, may be constrained by the choice of ticket sales model and the choice of use of a future stadium.

Finally, the Roland Garros case, in the same perspective as the BNPPM, associates reputation and physical resources as a resource base within the BM. The potential for controlling resources is associated with the potential of the Roland Garros stadium which will be modernized and which will develop towards this aim. In the case of Roland Garros, dependence relationships between the tournament and certain stakeholders and providers of associated resources are reversed. Certain partners and media become dependent on RG at the centre of their own business plan. As is underlined by Amis (2003), certain sporting events become strategic events in themselves for certain sponsors and the power of control thus finds itself reversed. The rank, the history and the strategic choices made under the leadership of Jean-Claude Blanc register in this perspective and make RG an illustration of an intentional external control strategy (Pfeffer & Salancik, 1978) for certain partners. The scope of control thus depends on the configuration of the event business model and its resource base.

Questions of the long-term survival and of the development of each of the organizations studied refer back to the power of the decision-makers to invest and develop

[16] As a reminder, in the tournament category to which the Open13 belongs, organizers have the possibility to give out financial guarantees to attract a player. In general, the factors which attract the best players, for this type of tournament, are the amount of the participation guarantee paid and the date of the event. The latter plays a strategic role in terms a player's sporting preparation and his or her strategic choices for the arrangement of his or her tennis season in terms of preparation tournaments before the major competitions (Masters Series and Grand Slam).

[17] Before being appointed to the head of the BNPPM, the choice of the tournament's pivot resource was associated with the main partnerships.

[18] For the 2008 event, over 100,000 seats were soled with an optimal commercial use of the potential of the POPB in terms of mass audience entertainment.

their business model around certain resources (Holcomb et al., 2009). The latter must not only have a significant individual potential, but also a synergy potential (Haspeslagh & Jemison, 1991) which eventually enables the transformation of the environment rules and the control of certain stakeholders who are significant contributors in terms of resources (Pfeffer & Salancik, 1978) by making them potentially dependent on the event.

The environmental instability which fragilizes the control exercised over the key resources mainly affects the reputation of events through the sporting line-up (participation then performance of high level sportsmen and women). The organizer does not control the sporting sphere and will seek to limit their dependence on the suppliers of sporting events who are the players. The three BMs analyzed suggest three different but coherent responses in relation to the choice of management of the event resources. For Open13, the management of the tournament relies on the financial resources generated by the relationship and partnership potential of the event, combined with the individual abilities of its director to attract the best athletes of the moment. For the BNPPM, the participation of the best players of the moment is only possible through the event maintaining its rank[19] in association with a specific unit dedicated to the monitoring of the players throughout the year on the professional circuit. Finally, in an inverse rationale, RG, through its status as a Grand Slam tournament, has the capacity to create the reputation of a player and so to make sportsmen dependent on the event. The necessary condition to maintain this status is the ability to make the tournament develop towards excellence in terms of venue, of direct[20] and indirect[21] exposure in comparison with the other Grand Slam tournaments or events which may take place in the future.[22]

5 Conclusion

The prolongation of the RBV trend of thought based on the configuration and the orchestration of assets have recently been put forward (Sirmon et al., 2009; Holcomb et al., 2009) while including the question of the putting in touch of resources (Dyer & Singh, 1998) as much in terms of their interconnection (Black & Boal, 1994) and of their co-specializations (Teece, 2007) as of the contributions of the stakeholders (Coff, 1999; Rindova & Fombrun, 1999) in relation to the decision unit studied.

In this perspective, the BM can appear as an integrating concept in terms of intentional choices of resource management (Tapscott, 2001; Hamel, 2002) and

---

[19] From the Association of Tennis Professionals (ATP) standpoint, maintaining it requires a minimum attendance (100,000 spectators) and also a certain amount (€2,270,000) of prizes ("Prize Money") awarded to the players which is expected to increase by €700,000 next year.

[20] Seating capacity in the stadium

[21] Media exposure (TV, radio, internet, press)

[22] For the next season, the date and the court surface of the Madrid tournament (Master Series category) will change to take place in April in a new stadium whose capacity is greater than that of Roland Garros.

allow the taking into account of the manner in which the focus organization, the unit of command analyzed, is linked to its stakeholders (Amit & Zott, 2001).

The main contribution of our research relates to the integrating nature of the BM concept in the RBV trend of thought, based on the management of the firm's resources (Mahoney, 1995; Sirmon et al., 2007, 2009; Holcomb et al., 2009). In this case, this is considered as the result of an empirical sequential approach to the construction and development of a portfolio of assets (Morgan, 2000; Sirmon et al., 2007) considered as a unique integrated command system (Penrose, 1959).

The central point of the approach concerns the firm's choice of asset management. Our main conclusion is that these choices are guided by a more or less explicit desire to take control of certain external elements of the organization (Pfeffer & Salancik, 1978; Freiling, 2008) and the putting in a dependency situation of certain stakeholder contributors and co-builders in terms of resources.

In the same tradition as Sirmon et al. (2009); Holcomb et al. (2009), but according to a different methodological approach, the ground covered has allowed direct and participative observation of these phenomena but the analysis remains limited to a type of industry where the competitive environment has little impact on decision making.[23] Moreover, only the aspects linked to revenue generation have been taken into account, both in terms of the RBV approach and the formalizing of the BM. The evaluation of the individual and collective potential of the resource panel is only suggested here.

The approach thus deserves to be broadened and to be registered in an analysis of performance factors in an evolutionist perspective (Durand, 2006), as well as in an evaluation of the contribution of certain stakeholders and of the aptitude of an organization to control its key resources in an uncertain situation.

References

Amis, J. (2003). Good things come to those wait: The strategic management of image and reputation at Guiness. *European Sport Management Quarterly, 3*(3), 189–214.

Amit, R., & Schoemaker, P. J. H. (1993). Strategic assets and organizational rent. *Strategic Management Journal, 14*, 33–46.

Amit, R., & Zott, C. (2001). Value creation in e-business. *Strategic Management Journal, 22*, 493–520.

Barney, J. B. (1991). Firm resources and sustained competitive advantage. *Journal of Management, 17*, 99–120.

Bingham, C. B., & Eisenhardt, K. M. (2008). Position, leverage an opportunity: A typology of strategic logics linking resources with competitive advantage. *Management and Decision Economics, 29*, 241–256.

Black, J., & Boal, K. (1994). Strategic resources: Traits, configurations and paths to sustainable competitive advantage. *Strategic Management Journal, 15*, 131–148.

Blyler, M., & Coff, R. W. (2003). Dynamic capabilities, social capital, and rent appropriation: Ties that split pies. *Strategic Management Journal, 24*(7), 677–686.

[23] Organizers of sporting or popular events often tend to say that their main competition was their own event the previous year.

Chesbrough, H., & Rosembloom, R. S. (2002). *The role of the business model in capturing value from innovation: Evidence from Xerox corporation's Technology Spinoff Companies* (Working Paper). Harvard Business School, 01-002.
Coff, R. W. (1999). When competitive advantage doesn't lead to performance: The resource-based view and stakeholder bargaining power. *Organization Science, 10*(2), 119–133.
Durand, R. (2006). *Organizational evolution and strategic management*. London: Sage Publications.
Dyer, J. H., & Singh, H. (1998). The relational view: Cooperative strategy and sources of interorganizational competitive advantage. *Academy of Management Review, 23*(4), 660–679.
Eisenhardt, K. M. (1989). Building theories from case study research. *Academy of Management Review, 14*(4), 532–550.
Eisenhardt, K. M., & Martin, J. A. (2000). Dynamic capabilities: What are they? *Strategic Management Journal, 21*, 1105–1121.
Fombrun, C. J., & Van Riel, C. B. M. (2004). *Fame fortune. How successful companies build winning reputations*. New York: Prentice Hall.
Freiling, J. (2008). RBV and the road to the control of external organizations. *Management Revue, 19*(1–2), 33–52.
Godfrey, P. C., & Hill, C. W. L. (1995). The problem of unobservables in strategic management research. *Strategic Management Journal, 16*, 519–533.
Granovetter, M. S. (1973). The strength of weak ties. *American Journal of Sociology, 78*, 1360–1380.
Grant, R. M. (1991). Resource-based theory of competitive advantage: Implication for strategy formulation. *California Management Review, 33*(3), 114–135.
Greenwood, D. J., & Levin, M. (1998). *Introduction to action research, social research for social change*. London: Sage.
Hamel, G. (2002). *Leading the revolution*. Boston: Harvard Business School Press.
Haspeslagh, P. C., & Jemison, D. B. (1991). *Managing acquisitions – Creating value through corporate renewal*. New York: Free Press.
Hayward, M. L. A., Rindova, V. P., & Pollock, T. G. (2004). Believing one's own press: The causes and consequences of CEO celebrity. *Strategic Management Journal, 25*, 637–653.
Helfat, C. E. (1997). Know-how and asset complementarity and dynamic capability accumulation: The case of R&D. *Strategic Management Journal, 18*, 339–360.
Henderson, R. M., & Cockburn, I. (1994). Measuring competence? Exploring firm effects in pharmaceutical research. *Strategic Management Journal, 18*, 5–14.
Holcomb, T. R., Holmes, R. M., & Connely, B. L. (2009). Making the most of what you have: Managerial ability as a source of resource value creation. *Strategic Management Journal, 30*, 457–485.
Kor, Y. Y., & Mahoney, J. T. (2005). How dynamics, management, and governance of resource deployments influence firm-level performance. *Strategic Management Journal, 26*(5), 489–496.
Lincoln, Y. S., & Guba, E. G. (1985). *Naturalistic inquiry*. London: Sage.
Lippman, S. A., & Rumelt, R. P. (2003). A bargaining perspective on resource advantage. *Strategic Management Journal, 24*(11), 1069–1086.
Magretta, J. (2002). Why business matter. *Harvard Business Review, 80*(5), 3–8.
Mahoney, J. T. (1995). The management of resources and the resource of management. *Journal of Business Research, 33*, 91–101.
Miller, D., & Shamsie, J. (1996). The resource-based view of the firm in two environments: The Hollywood film studios from 1936 to 1965. *Academy of Management Journal, 39*, 519–543.
Morgan, R. M. (2000). Relationship marketing and marketing strategy: The evolution of relationship strategy within the organization. In J. Sheth & A. Parvatiyar (Eds.), *Handbook of relationship marketing* (pp. 481–504). Thousand Oaks, CA: Sage Publications.
Parent, M. M., & Foreman, P. O. (2007). Organizational image and identity management in large-scale sporting events. *Journal of Sport Management, 21*, 15–40.
Penrose, E. (1959). *The theory of the growth of the firm*. Oxford: Basil Blackwell.

Peteraf, M. A. (1993). The cornerstones of competitive advantage: A resource-based view. *Strategic Management Journal, 14*, 179–191.

Pfeffer, J., & Salancik, G. (1978). *The external control of organization – A resource dependence perspective*. Stanford: Stanford University Press, 2003. Originally published: New York: Harper & Row.

Porter, M. E. (2001, March). Strategy and the internet. *Harvard Business Review, 79*, 63–78.

Prahalad, C. K., & Hamel, G. (1990, May-June). The core competence of the corporation. *Harvard Business Review, 68*, 79–91.

Priem, R. L., & Butler, J. E. (2001). Is the resource-based "view" a useful perspective for strategic management research? *Academy of Management Journal, 26*(1), 22–40.

Reed, R., & DeFilippi, R. J. (1990). Causal ambiguity, barriers to imitation and sustainable competitive advantage. *Academy of Management Review, 15*(1), 88–102.

Rindova, V. P., & Fombrun, C. J. (1999). Constructing competitive advantage: The role of firm-constituent interactions. *Strategic Management Journal, 20*, 691–710.

Rouse, M. J., & Daellenbach, U. S. (1999). Rethinking research methods for the resource-based perspective: Isolating sources of sustainable competitive advantage. *Strategic Management Journal, 20*, 487–494.

Rouse, M. J., & Daellenbach, U. S. (2002). More thinking research methods for the resource-based perspective. *Strategic Management Journal, 23*, 963–967.

Rumelt, R. P. (1984). Towards a strategic theory of the firm. In R. Lamb (Ed.), *Competitive strategic management* (pp. 556–570). Englewood Cliffs, NY: Prentice-Hall.

Schultz, M., Hatch, M. J., & Larsen, M. H. (2000). *The expressive organization. Linking identity, reputation and the corporate brand*. Oxford: Oxford University Press.

Shafer, S. M., Smith, J. H., & Linder, J. C. (2005). The power of business models. *Business Horizons, 48*, 199–207.

Sirmon, D. J., Gove, S., & Hitt, M. A. (2009). Resource management in dyadic competitive rivalry: The effect of resources bundling and deployment. *Academy of Management Journal, 51*(5), 919–935.

Sirmon, D. J., Hitt, M. A., & Ireland, R. D. (2007). Managing firm resources in dynamic environments to create value: Looking inside the black box. *Academy of Management Review, 32*(1), 273–292.

Srivastava, R. K., Fahey, L., & Christensen, H. K. (2001). The resource-based view and marketing: The role of market-based assets in gaining competitive advantage. *Journal of Management, 27*(6), 777–802.

Srivastava, R. K., Shervani, T. A., & Fahey, L. (1998). Market-based assets and shareholder value: A framework for analysis. *Journal of Marketing, 62*(1), 2–18.

Strauss, A. L., & Corbin, J. (1998). *Basics qualitative research techniques and procedures for developing grounded theory* (2nd ed.). Thousand Oaks, CA: Sage Publications.

Tapscott, D. (2001). Rethinking strategy in a networked world. *Strategy+Business, 24*, 1–8.

Teece, D. J. (2007). Explicating dynamic capabilities: The nature and microfoundations of (sustainable) enterprise performance. *Strategic Management Journal, 28*, 1319–1350.

Teece, D. J., Pisano, G., & Shuen, A. (1997). Dynamic capabilities and strategic management. *Strategic Management Journal, 18*(7), 509–533.

Timmers, P. (1998). Business models for electronic markets. *Electronic Market, 8*(2), 2–8.

Venkatraman, N., & Henderson, J. C. (1998). Real strategies for virtual organizing. *Sloan Management Review, 40*(1), 33–48.

Wernerfelt, B. (1984). A resource-based view of the firm. *Strategic Management Journal, 5*, 171–180.

Whyte, W. F. (1991). *Participatory action research*. London: Sage.

Yin, R. K. (1994). *Case study research: Design and methods* (2nd ed.). Applied social research method series: Vol. 5. Beverly Hills, CA: Sage Publications.

Zott, C., & Amit, R. (2007). Business model design and the performance of entrepreneurial firms. *Organization Science, 18*(2), 181–199.

Governance and Sporting Success of Top 20 Football Clubs After Economic Crisis

Domenico Marino

Abstract We study the relationship between the governance and sporting success in team sports. A neutral network model is developed and applied to the top 20 European football clubs with the aim of determine those teams that have a greater potential for sporting success in the years to come. The proposed approach also allows to identify those factors that are crucial for sporting success.

1 Introduction

The arguments over the relationship between sporting success and economic variables have gone on for years. The case of football is surely the most interesting and even more important example from a quantitative point of view. Similar things could be said about other sports such as Formula 1 for example. The issue of sporting success is not simply an aspect that is linked to the epic and immeasurable size of a sporting talent, which nevertheless still remains an inescapable element to success, but it must take its place alongside other issues such as the governance of clubs, investments, the system of regulations, the structure of a club's costs and profits.

What we are referring to obviously concerns team sports, sports in which results do not depend on an individual's performance, but becomes the sum total of so many individual performances, even if in some cases with differing roles and responsibilities. Sporting success therefore presents itself with a collective connotation and contributes to the increase of a team's long term value.

2 The European Market

Deloitte's 2009 report brings to light some interesting facts such as the European football market being worth € 14.6 billion.

D. Marino (✉)
Mediterranea University of Reggio Calabria, Reggio Di Calabria, Italy
e-mail: dmarino@unirc.it

S. Butenko et al. (eds.), *Optimal Strategies in Sports Economics and Management*, DOI 10.1007/978-3-642-13205-6_5,

The five biggest football federations (the Big Five) alone control some 50% of the market with a turnover of € 7.7 billion, an increase of 10% over the previous year. Of these *Big Five* it is the Premier League that tops the list having generated revenues of € 2.4 billion taking advantage of the new agreements on television rights and leaving its foreign competitors lagging even further behind. This is something that becomes even more apparent with the exchange rate for Euro—Sterling. Spanish teams Barcelona and Real Madrid have recorded notable hikes in revenue, making them the third and first teams in the world for revenue volumes. These increases have meant that the Spanish Liga is now on an equal footing with the Bundesliga with € 1,438 billion pushing it above Italy's Serie A that is next in line with € 1,428 billion. Propping up the *Big Five* is French football that has recorded revenue volumes in the region of € 1 billion. Of the five it is the Italian federation that has seen the greatest rise in its revenues with an increase of 34% shared almost equally between three types of revenue sources: merchandising, television rights and match day attendances. An analysis of the different teams shows that these percentages differ totally, with the Spanish teams and Bayer focussing on merchandising, the Italians on television rights and the English on stadium ticket sales. In general football shows itself to be a counter-cyclical sector that is growing even in the midst of an economic crisis.

There are two fundamental aspects that characterize this sector: illiquidity and intangibility. A sports club's capital is essentially made up of the value of its athletes. That value is subject to severe fluctuation, caused by the physical condition and performance of the athletes. Furthermore, assessing that value is like assessing any other intangible asset. The product—the match played—is in itself an intangible asset. These two characteristics give us a little insight into some of the problems that have arisen in recent years in sports management.

3 The Governance of Sporting Activities

There is no doubt that the problem of the governance of sporting clubs and those of football clubs in particular is a central issue. This can basically be split into the following three different types of governance: entrepreneurial model of governance; association-type model of governance; patronage-based model of governance. The entrepreneurial model and patronage-based model are those found mainly in Italy and England, whilst the association-type model constitutes the backbone of the Spanish system. The limitations of the entrepreneurial system are clear to see and are associated with: (a) limited return on investments; (b) the irrelevance of the club's capital and its easy depreciation; (c) high running costs. The patronage based model is threatening to intoxicate the market because it allows some to invest without taking into account the returns on their actual investments and for objectives that are totally set apart from sport itself. Berlusconi's Milan is in this case very symptomatic of this. The association-type model appears to be the one that works the best if we consider that Real Madrid and Barcelona are among the first three clubs in the world in 2008 for revenue, and they operate on an association-type model. Nevertheless this model is too closely linked to specific territorial issues and is not easily exportable or generalizable. The football market therefore is also characterized by the confronting of these different operating models that are often decisive in sporting success or even failure. Buying and selling players and the resultant strengthening or weakening of a team find different meanings and outcomes in the various operating models. The strengthening of a team is a way of boosting an image in the case of the patronage-type or contributing towards the election of a chairman in the case of the association-type. In the entrepreneurial model the weakening of a team can be an attempt to wipe out debts or eliminate annual losses.

Financial fair play in this context manifests itself as a regulatory mechanism that attempts to wipe out this starting point disparity, imposing a cap on spending and avoiding anyone splashing out extravagantly. The advantages of this system are greater long-term financial stability, less debt and a greater attention paid to the managing of the breeding grounds. The price that will have to be paid for these advantages will be in terms of the entertainment, for we will no longer see "star-studded" teams.

4 Baumol's Disease and a Simple Financial Fairplay Rule

The football sector appears to be suffering from a particular form of what is known as Baumol's disease. In 1966, Baumol and Bowen drew attention to the fact that certain economic sectors, including that of the performing arts and healthcare sector are subject to continuous increases in costs. This is due to the fact that productivity in these sectors does not increase or rather, it increases very little by comparison to other sectors in any given period. If we include the entertainment sector as one of those subject to Baumol's disease, to all purposes sport, and specifically football, proves to be subject to this particular phenomenon. The effect of Baumol's disease however, is different from the one displayed in the culture-related sectors. Negative growth in productivity, which in the case of cultural activities is caused by the rising costs of repeating the performance, in the case of sports-related activities is accentuated by the fact that competition between the various clubs, together with the need to achieve certain results, imposes higher and higher costs on the clubs. Additionally, the need to ensure the involvement of champions causes the number of signings to rise. These reflections allow us to guess that the mainly financial recession that the football industry as a whole is currently going through is not an incidental factor, but rather a structural feature of the system, which in the absence of external corrective action will no doubt evolve towards bankruptcy. In this scenario, rules like financial fair play take on a new and important role. It is not a question of rules to correct a specific incidental state of affairs, but rather a way to correct the structure of the sector itself. Since any form of public subsidy of the sector is ruled out a priori, given that by comparison to other sectors affected by Baumol's disease, such as culture and healthcare, it generates a low level of social capital, the solution to Baumol's disease must be sought in market mechanisms. Based on this, financial fair play is doubtless an attractive regulating mechanism. The option to exploit television rights is a special factor within the sector and to all intents and purposes can play an important role in substituting public funding or private donations. The system worked as long as the fees paid for television rights increased year by year. This created the illusion that the continuous growth in revenue generated by the rights would cover any management disasters. However, this growth could not be sustained and when the fees stopped increasing and in some cases began to decrease, the football industry as a whole went into recession. In order to find a simple fair play rule, we need to try to create a model for the basis of football club profits. We can express this using the formula below:

$$\Pi = R - C - G + S - M. \tag{1}$$

Let R be total revenue, C costs linked to signings, S the non-operating incomes deriving from trading players, M the depreciation in the value of the players, and G general costs. Hence a simple fair play rule may be expressed as follows:

$$R \geq C + G \text{ and } S \geq M. \tag{2}$$

This means that the clubs must abide by two rules, one linked to revenue and signing costs and the other to investments. One of these rules may be temporarily broken if the sum

$$R - C - G + S - M \geq 0. \tag{3}$$

In this manner, the long-term stability of football clubs would be guaranteed and the clubs would have incentive to invest in academies and harness internal efficiency. If this were to occur, sporting success would be the result of proper and efficient management, which can only be an advantage to the game.

5 Neural Networks and the Problem of Predicting Sporting Success

Neural networks and architecture constitute one of the more recent and most interesting fields of research, fields that began originally with the study and research of artificial intelligence, but gradually spread to include other sciences. The approach based on "neural networks" is referred to as connectionist and allows us to think about the phenomena of learning and memory, as will be seen further on. The early concepts of this new science, known as neuroinformatics, are attributable to McCulloch and Pitts and the fundamental idea behind their work was to attempt to reproduce the cerebral structure, founded on neurons, axons and dendrites using a mathematical model. Every neuron receives electrical impulses from dendrites and retransmits them via axons. It is therefore possible, in principle, to imagine constructing a minute model of a cerebral structure using logical portals, easily describable using the rules of Boolean algebra. The neuron, therefore, is modeled as a logic portal that has the goal of modifying and retransmitting the signal, obeying a specific rule. The transmission of the signal depends on several parameters: on the internal status of the logic portal, on the status of the previous portals and the signal received. The output is determined by means of an activation function that can be linear, non linear or threshold and that depends on the type of decision that the system has to simulate or logical operations that the system is called upon to complete. A quick comparison with the cerebral structure reveals that the human brain has 100 billion neurons and each neuron approximately 10,000 synapses or connections. The brain's extraordinary capabilities are down to this enormous number of connections. Each neuron receives and processes information coming from other neurons producing an outgoing impulse that is sent, via the synapses, to the connected neurons. Synapses can be excitatory or inhibitory and every neuron is activated or inhibited depending on whether the overall stimulation does or does not exceed a certain threshold. An artificial neuron structure consists in a collection of nodes that make up the units of logical calculus (in a Boolean sense), connected by communication channels (synapses), that constitute the structure for storing data. Each neuron can have a g threshold and an s status. Synapses have a synaptic weight associated with them. The system has three dynamic characteristics: activation, learning and iteration.

The first updates the status of the neurons, the second alters the weight of the connections and the third regulates the order of activation. A neural network is capable of learning a prearranged model. It is said that to do this the network is trained so that the introduction of an input can result in a desired output or one that is consistent with the same.

The training consists of presenting the neural network with a combination of inputs and, as a result, modifying the synaptic weights in such a way to obtain the considered outputs. The combination of models presented to the network during the learning phase is defined as a training set. The models used to confirm the level of learning are contained in the *validation set.* The learning algorithms can be divided into two groups:

(a) supervised;
(b) unsupervised.

The first requires each input model to be associated with a desired output model. The training set therefore contains a lot of information about each of them. The network produces a calculated output for each model presented that is compared with that desired. Any error is transmitted back along the network, correcting the synaptic weights in accordance with a learning algorithm that tends to minimize the mentioned error. The training of the network comes to an end when a level is achieved for the calculated error that is compatible with a certain confidence interval previously established.

Unsupervised algorithms within the training set, on the other hand, contain only input models. Learning algorithms modify the synaptic weights to produce output carriers that are consistent. The training process therefore extracts the statistical properties of the training set and groups similar models into similar classes. An interesting feature of neural networks is that of being able to discern and extract a model of a general nature, even where it is distorted by noises. The neural networks can also adapt flexibly to situations that are complex and changeable in time.

The EBP (Error Back Propagation) learning algorithm is the best known and the fastest learning algorithm for supervised networks. It has a multi-layered structure, that is to say it has hidden layers. Each neurone is connected with that of the previous layers without horizontal connections. The EBP algorithm has two phases. In the first the signal is transmitted from the exterior towards the interior. In the second the direction is reversed.

A defined error can be calculated:

$$E_{\pi} = \frac{1}{2}\sum(t_{\pi j} - \sigma_{\pi j}), \tag{4}$$

where $t_{\pi j}$ and $\sigma_{\pi j}$ indicate the anticipated output and the calculated output of the j-th neuron, respectively. The output calculated for a generic neuron is the function of the contribution of the neurons connected with it:

$$\sigma_{\pi j} = f_j(\mathrm{net}_{\pi j}) \tag{5}$$

with f generally assuming the shape of a sigmoid. Here $\mathrm{net}_{\pi j}$ represents the contribution of the associated neurons weighted with synaptic weights:

$$\mathrm{net}_{\pi j} = \sum_i \pi_{ij}\sigma_{\pi j} \tag{6}$$

The second phase consists in the correction of the synaptic weights with the aim of minimizing the error. Updating of synaptic weights follows the following criterion:

$$\varDelta p = \varepsilon\delta_{\pi j}\sigma_{\pi j} \tag{7}$$

with ε defined as the level of learning.

The application of neural networks in finance affects a large range of problems, some of which are extremely complex. The capacity and the characteristics of the networks allow a whole series of applications, from the simplest to the most complex. An interesting aspect is that associated with classification. The classification properties of a neural network constitute one of the richest fields for possible applications. By comparing this with the traditional methods of classification (*Discriminant Analysis*), commonly used in financial statistics, it is possible to show quite clearly how the use of neural networks results in greater levels of efficiency.

Table 1 Some financial indicators of Top 20 football clubs

Club	Rev. 2006 (10^6 €)	Rev. 2007 (10^6 €)	Rev. 2008 (10^6 €)	Oper. income (10^6 \$)	Nursery	Debt/Value	% Rev. Match day	% Rev. Mass media	% Rev. Merchandise	Current value (10^6 \$)	1-Year % value change
Real Madrid	276	366	351	81	10	23	23	32	45	1353	5
Manchester United	246	325	315	160	10	54	42	29	29	1870	4
Barcelona	208	308	290	108	10	7	32	38	30	960	22
Chelsea	221	269	283	−13	8	92	38	37	25	800	5
Arsenal	171	264	264	80	8	107	32	42	26	1,200	0
Milan	234	210	227	58	8	0	16	59	25	990	24
Bayern Munich	190	295	223	59	8	0	38	0	62	1,110	21
Liverpool	181	211	199	50	9	59	27	42	31	1,010	−4
Inter	177	173	195	27	9	77	20	58	22	370	−8
Roma	132	175	158	69	8	9	21	58	21	381	−12
Tottenham	104	145	153	70	8	29	30	36	34	445	8
Juventus	229	168	145	46	9	5	10	54	36	600	18
Olympique Lyonnais	93	156	141	94	8	18	22	49	29	423	4
Newcastle	129	126	129	−13	8	96	41	32	27	285	−5
Hamburg	79	128	120	44	7	0	40	20	40	330	13
Schalke 04	97	148	114	41	7	38	24	17	59	510	9
Celtic	93	104	112	11	8	14	50	27	23	218	−4
Valencia	85	104	108	0	8	0	28	53	19	204	0
Olympique Marseille	79	104	99	20	7	0	30	40	30	240	28
Werder Bremen	79	104	97	27	6	0	40	20	40	292	12

The main aim of using neural networks in this work will be to determine, using the discriminatory capabilities of an appropriately trained neural network, those teams that have a greater potential for sporting success in the years to come. The indicator that describes sporting success is identified by using the level of growth of the value of the team as a whole. In order to obtain this result a database has been put compiled containing information on the financial aspects (costs, revenue, and investments), operational aspects and sporting aspects of the top twenty European teams. The database constitutes the knowledge base from which the neural network will extract its forecasts in the form of a discriminatory analysis. This will allow us to attempt to forecast future successes on the basis of given variables of a varied nature, not just sporting but also financial and operational. For our purposes the sporting success is defined as "increasing in time of current value of the club". This analysis will also allow us, using the neural network, to identify those factors that are crucial for sporting success. We will in particular endeavor to respond to the question as to whether models based on "financial fair play" are effective and whether financial sustainability will become a limiter to both results and entertainment.

Table 1 summarizes the values of some variables that are the subject of the analysis:

6 The Results of the Neural Classification and Conclusion

Table 2 highlights the results of the neural network forecasts. As can be clearly seen, today's top teams shows lower growth levels compared to the up and coming teams, as though a catching up mechanism has kicked in. These results are very

Table 2 Neural network forecasts of future 1-year value change, %

Club	Change, %
Real Madrid	5.0
Manchester United	4.7
Barcelona	5.8
Chelsea	6.6
Arsenal	7.3
Milan	7.9
Bayern Monaco	2.9
Liverpool	8.5
Inter	10.4
Roma	5.1
Tottenham	7.7
Juventus	7.6
Olympique Lyonnais	8.3
Newcastle	7.7
Hamburg	6.6
Schalke 04	6.0
Celtic	6.1
Valencia	7.7
Olympique Marseille	9.1
Werder Bremen	8.2

interesting. The first consideration concerns the so called "financial fair play". The neural network analysis shows that financial sustainability of football clubs will not become a limiter to both results and entertainment, but will be an incentive to invest in nursery and to program the long-term development. It is easy to win by purchasing the best champions. But this policy is more expensive and not always effective. It is a myopic policy. The football market needs a prospective policy to achieve a long run success. The second consideration concerns the critical factors of success in football market. A correct management of resources and a good technical competence is more important than the "intensity" of capital. In a period of financial crisis in all the world and in all the sectors this aspect is very important!

References

Baumol, W.J. (1993). Health care, education, and the cost disease: A looming crisis. *Public Choice, 77*, 17–28.

Baumol, W. J., & Bowen, W. G. (1966). *Performing arts. The economic dilemma*, Cambridge, MA: Twentieth Century Found.

Deloitte & Touche (2007). *Deloitte Football Money League 2007*, London.

Deloitte & Touche (2008). *Deloitte Football Money League 2008*, London.

Deloitte & Touche (2009). *Deloitte Football Money League 2009*, London.

Gil Lafuente, J. (2002). *Algoritmos para la excelencia*. Claves para el éxito en la gestiốn deportiva. Vigo: Milladoiro.

Marino, D. (1998). *Dinamica Economica, Complessitáe Caos*. Napoli: Liguori.

Szymanski, S. (1999). Why is Manchester United so successful? *Business Strategy Review, 9*(4), 47–54.

The Stadium Game in an Uncertain Environment: A Preliminary Look at Arena Discourse in Edmonton, Canada

Daniel S. Mason

Abstract Stadiums and arenas, built as anchors of broader urban development projects, have become a flashpoint of debate among taxpayers and scholars alike. This is because the attachment that citizens have with their local sports teams, combined with restraints on the number of available franchises, has resulted in these developments becoming the most substantively funded pieces of the urban development puzzle, from a taxpayer perspective. This chapter provides a preliminary examination of media discourse in a Canadian city currently considering the construction of an arena in its downtown core. A discussion of four frames identified by Buist and Mason (in press): (1) economic development, (2) civic status, (3) civic priorities, and (4) financing, is provided. In addition, given the current state of the global and regional economies, data were also coded for an additional frame—economic downturn. In the second phase of data analysis, an inductive approach was undertaken to determine if any additional frames existed in the Edmonton context, which resulted in the identification of frames relating to the comprehensiveness of the project and the need for public consultation. The examination of newspaper coverage in Edmonton over the past year has revealed that many of the same frames found in other debates over the public subsidization of sports facilities in other cities are present in the current context. In the case of Edmonton, because the process is ongoing and has been drawn out over time, additional speculation regarding the financing of the project has become a dominant frame. Interestingly, bold claims of economic impacts have not been used by proponents to the same degree as elsewhere; however, as the project moves forward this does not preclude some of these claims from being made. It is important to note that this chapter has only focused on a portion of the coverage of an ongoing issue. If and when the issue is resolved in Edmonton, a content analysis of the entire body of newspaper coverage might yield more insights into the balance found of media coverage.

D.S. Mason (✉)
Faculty of Physical Education and Recreation, University of Alberta, Edmonton, AB, Canada T6G 2H9
e-mail: dmason@ualberta.ca

S. Butenko et al. (eds.), *Optimal Strategies in Sports Economics and Management*, DOI 10.1007/978-3-642-13205-6_6, © Springer-Verlag Berlin Heidelberg 2010

1 Introduction

One unique characteristic of North American major league sports franchises relates to their mobility—they are businesses that can and do uproot and seek out better opportunities elsewhere where local communities fail to support them. However, due to the unique treatment of sports franchises by antitrust and competition laws—which allow leagues to limit the number of available franchises—these same sports franchises have also developed a means to leverage the presence of a local team to exact significant subsidies from their host communities; thus, rather than moving away from financially untenable situations, teams may move from good financial situations to even greater ones, all at the expense of local taxpayers.

There are several reasons why cities continue to engage in the "stadium game." The first relates to the competitiveness of cities. Due to the changing global economy, cities now compete on a much broader scale for flows of capital, and many have been forced to reinvent themselves in the context of a movement away from traditional industries—such as manufacturing and distribution—toward a more service based economy (Sassen, 2002). Thus, as human capital becomes more mobile, cities try to present themselves as attractive places to visit, work, and invest in. To do so, significant public resources have been invested in convention centres, the construction of downtown "tourism bubbles", and the development of amenities that contribute to tourism and quality of life benefits (Spirou & Bennett, 2003).

Within this competition, stadiums and arenas, built as anchors of broader urban development (read revitalization) projects have become a flashpoint of debate among taxpayers and scholars alike (Danielson, 1997; Euchner, 1993; Rosentraub, 1997a; Shropshire, 1995). This is because the attachment that citizens have with their local sports teams, combined with restraints on the number of available franchises, has resulted in these developments becoming the most substantively funded pieces of the urban development puzzle, from a taxpayer perspective. And like other pro-growth development projects, there are those who stand to benefit significantly from sports facility development.

Academic research has examined the stadium game from a number of different lenses. The first relates directly to the financial benefits that teams generate for their host communities. Perhaps in response to over-inflated economic impact studies put forth by facility proponents, a small group of academics began exploring the economic benefits of teams. This body of research has generally debunked the myth that sports arenas and their member franchises act as engines of economic growth within communities and, where measurable, economic impacts have even proved to be negative (cf. Baade & Dye, 1988, 1990; Baim, 1994; Coates & Humphreys, 2000, 2003; Colclough Daellenbach & Sherony, 1994; Fainstein & Stokes, 1990; Hudson, 1999; Humphreys, 2001; Rosentraub, 1997a, 1997b; Saidel, 2000; Zipp, 1996).

Another stream of research has been lead by economists seeking to measure the intangible benefits that teams have. These studies have attempted to ascertain the public goods value of teams as a way of evaluating the use of public subsidies. While this research has confirmed that there are indeed significant public goods associated with local sports franchises, they do not justify the size of subsidies that teams have received (cf. Johnson, Groothuis, & Whitehead, 2001; Johnson, Mondello, & Whitehead, 2007; Santo, 2007).

Other scholars have sought to examine just how cities come to make the decision to subsidize teams, borrowing from urban affairs and political economic approaches. Thus, a focus is on the decision making apparatus and how facility proponents are able to get their way in providing the subsidy (Friedman & Mason, 2004, 2005). Further, other scholars have examined the role that the media play in shaping public opinion, in light of the fact that the media are not disinterested parties in regard to the issue (Delaney & Eckstein, 2008). Buist and Mason (in press) examined how the local newspaper framed coverage of two separate proposals to build sports facilities using public funds in Cleveland, Ohio; one that failed in 1984, and another that passed in 1990. They found that several distinct frames were embedded in discourse surround coverage in the Cleveland *Plain Dealer.* As the discussion above would suggest, it is critical to better understand exactly how discourses surrounding facility development in communities emerge, as these provide the messages that shape public opinion. Clearly more empirical research is required to examine this issue in practice. The following extends this work by providing a preliminary examination of media discourse in a Canadian city currently considering the construction of an arena in its downtown core. The paper is organized as follows. First, an overview of the stadium game in North America is provided, followed by a description of the context in Edmonton, Alberta. Following a review of the method, the media frames emerging from the newspaper analysis are discussed.

2 An Overview of the Stadium Game

As mentioned above, decisions to fund sports facilities are ultimately rooted in the rivalries amongst cities competing in globalizing economies for tourists, jobs, and investment (Euchner, 1993; Harvey, 1989; Kotler, Haider, & Rein, 1993; Waitt, 1999). Sports facilities are viewed as particularly useful in cities where there is a transformation from traditional industries to more service-based economies (Laws, Scott, & Parfitt, 2002; Whitelegg, 2000), and where destinations are developed in urban areas for local and visitor consumption (Hannigan, 1998). The perception of increasing competition amongst cities has also created a sense of urgency for some communities (Altshuler & Luberoff, 2003), igniting a stadium and arena construction boom that has occurred in larger urban centers in North America, where 72 new facilities alone opened between 1990 and 2005. Costs have also increased dramatically—facilities that cost an average of US$51 million (in 2001 dollars) in the 1950s now exceed US$300 million (Altshuler & Luberoff, 2003). In addition, cities have chosen to use sports facilities as part of downtown redevelopment—79% of the new facilities built between 1990 and 2001 were situated in or very close to downtown, in contrast to 39% during the previous three decades (Altshuler & Luberoff, 2003). In sum, 80% of the professional sports facilities in the United States have been replaced or undergone major renovation since 1987, at a total cost of more than US$19 billion, with the public providing US$13.6 billion (71%), of that amount (Baade, 2003). Cities have traditionally paid for facilities using infrastructure and operating subsidies (Howard & Crompton, 2004), through general obligation and revenue bonds paid off using future stadium revenues, or hotel, admission, or sales taxes (Petersen, 2001).

In a study of minor league baseball stadiums, Johnson (2000) identified three objectives communities have in financing facilities. First, stadiums can be used to advance redevelopment strategies, such as anchoring downtown development. Thus, in return for funding the sports facility and providing a favorable lease agreement, cities expect certain positive economic benefits, including direct (visitor spending), indirect (ripple effects of recirculating dollars), and induced (further ripple effects of employees spending wages, etc.) effects (Howard & Crompton, 2004). In addition, cities anticipate that tax revenues will rise as a result of increased economic activity, or sharing in revenues generated by the venue itself (Rosentraub, 1997a). Second, a facility and team can enhance community image (cf. Baim, 1994; Duquette & Mason, 2008). Finally, sports facilities provide additional recreational amenities for residents and contribute to the community's quality of life. Examples of this might include "the satisfaction people get from living in a 'big league' town, from having another topic of conversation that is common to most citizens, [and] from reading about [a team's] successes and failures in the newspaper" (Zimmerman, 1997, p. 121). However, Johnson noted that "communities should not expect to recoup their stadium investment on a dollar-for-dollar basis" (Johnson, 2000, p. 146). This is because the actual economic development benefits of hosting franchises have been overstated (Coates & Humphreys, 2000, 2003; Hudson, 1999; Humphreys, 2001; Rosentraub, 1997a), so much so that enhanced image and quality of life arguments have become stronger arguments for facility proponents (Gratton & Taylor, 2000; Howard & Crompton, 2004; Shropshire, 1995). Nonetheless, economic arguments are still made in order to generate support locally for stadium construction, relying on faulty economic impact studies (Baim, 1994) "conducted by well-regarded consulting firms using generally accepted, though not necessarily accurate, techniques of analysis" (Utt, 1999, p. 4). For example, proponents of Toronto's Skydome (now Rogers Centre) promised that the facility would generate $450 million in its first year of operations and create 17,000 jobs (Coates & Humphreys, 2000). The stadium, which would ultimately cost taxpayers over $300 million (Rosentraub, 1997a), made $17 million in its first year of operations and was most recently sold in 2004 for $25 million (a fraction of the cost of its construction).

Once facilities are built, the interests of the city and the team quickly diverge. While teams are interested in moving to downtown locations in order to access lucrative corporate clients more easily (Altshuler & Luberoff, 2003), teams try to capture revenues solely within the facility, whereas cities hope for more spending and economic activity surrounding the facility (Gamrat, 2002). Other issues concerning stadium development include neglecting the opportunity costs of public investment, where alternative uses of funds are not considered (Noll & Zimbalist, 1997, p. 62), or the tendency for high-profile projects such as sports facilities to overshadow and/or marginalize other important civic concerns, such as schools, parks, housing, and libraries (Euchner, 1993).

Despite the apparent lack of benefits to cities described above, there are some who *do* stand to gain from the construction of sports facilities. In this case pro-growth advocates in cities continue to view a new sports facility as a solution to the economic development needs of their communities (Turner & Marichal, 2000). In other words, there are certain groups within cities who have an interest in ensuring that a facility is built. For example, franchises and local newspapers have a mutual interest in the home town, which leads papers to provide extensive and sympathetic coverage to local teams (Danielson, 1997; Friedman & Mason, 2004). Because newspapers and other media are often owned or controlled by part of the local power structure in a city, the media share a common interest with pro-growth advocates (Delaney & Eckstein, 2003). More specifically, "newspapers are expected to be sympathetic to the demands of other growth actors because the paper stands to profit from growth through circulation and advertising sales" (Turner & Marichal, 2000, p. 192). Thus, the media often provide a platform to promote a pro-growth agenda, which may include the construction of a sports facility. In addition, construction companies, contractors, banks and bankers who arrange for financing, architectural firms, and team owners also benefit (Zimbalist, 2003). Thus, while the cities themselves may not realize direct economic benefits, there is a concentrated group of stakeholders who stand to benefit from construction (Friedman & Mason, 2004). In contrast, those who stand to lose from stadium construction tend to be widely dispersed; "this means that proponents of public subsidies have a strong economic motivation to work very long and hard to ensure that the subsidies are forthcoming, but opponents have less incentive to resist the subsidies" (Rock, 2001, p. 4). It is also important to note here that stadium benefits are not limited to private interests, as political interests can be furthered by pushing a stadium agenda (by "saving" a team or acquiring a new one), or damaged by losing a team to relocation (Euchner, 1993; Riess, 2000). As a result, political and business elites that control decision making in a city have a strong incentive to work together to finance a sports facility (Howard & Crompton, 2004). With the construction of Jacobs Field in Cleveland, Sidlow and Henschen (1998) described the policy process as "an almost seamless web of wealthy, well-connected sports entrepreneurs, politicians, and business executives [who] were interested in continuing downtown revitalization, and saw a new stadium complex as part of the strategy" (p. 84).

3 A Proposed Arena Development Project in Edmonton

This chapter explores a contemporary example in the city of Edmonton, Alberta, Canada. The city of Edmonton has had a long standing affinity for the game of hockey, including successful amateur teams through the 1950s and 1960s. However, the city's first real taste of big-league hockey came when "Wild" Bill Hunter secured a franchise in the fledgling World Hockey Association (WHA) in 1972, which lead to the construction of a new arena in 1974. The facility was named Northlands

Coliseum (later Edmonton Coliseum, Skyreach Centre and finally Rexall Place), owned/controlled by Northlands, a non-profit entity that operates arm's length from the City.

The team had the good fortune of acquiring the rights to star player, Wayne Gretzky, and when the WHA eventually merged with the National Hockey League (NHL) in 1979, Gretzky remained with the Oilers. Drafting strong players to complement Gretzky, the team won its first Stanley Cup championship in 1984, only 5 years into the team's NHL existence. The team would win four titles in a 5 year period, and won again in 1990 after it had traded the rights to Gretzky to Los Angeles. Perhaps Edmonton hockey fans had been spoiled by the competitive success of the Oilers in the 1980s, and as the team embarked on a less successful series of campaigns in the early 1990s, attendance waned at games, and the limitations of the arena, which lacked many of the revenue generating amenities found in newer arenas, loomed larger. The arena underwent a significant renovation in 1994 that saw seating capacity drop in favor of the introduction of new luxury suites on the facility's concourse level.

Through 2005, the arena had undergone some periodic upgrades but by the fall of 2005, Oilers management considered the facility to be, in the words of its CEO, "getting tired" (Mah, 2007). Recognizing that the Oilers' lease with Rexall Place was expiring in 2014, the City undertook some preliminary research into arena development projects elsewhere in order to prepare the city for the team's impending expectations for a new facility. This occurred in the Fall of 2006 and was completed in the spring of 2007. In 2007, the City, Northlands, and the Oilers, under the guidance of the Edmonton Mayor, began a process to prepare a written document that provided some recommendations for a new or renovated facility for the Oilers. In the meantime, Northlands hired arena designers, HOK, to conduct a feasibility study related to the renovation of Rexall Place, which was completed in early 2007.

The Edmonton Arena Feasibility Committee was made up of prominent members of the community, including several with ties to the Oilers and Northlands. Two subcommittees were also formed, Financing and Community and Design. The group also commissioned a leading expert of sport and urban infrastructure development, Dr. Mark Rosentraub, who prepared a report as a supplement to the arena committee's final document, *City Shaping*, released in March of 2008. Among its recommendations, the document strongly advocated for locating the facility downtown, and using public funds, the debt which would be serviced by a "community revitalization levy" on business activities in the area surrounding the new arena.

In February of 2008, the University of Alberta and Edmonton Chamber of Commerce co-hosted a conference on The Role of Sports and Entertainment Facilities in Urban Development. The conference featured speakers from both the academic and business communities, and focused on the successes and challenges associated with broader, arena-anchored urban development projects in other cities throughout North America. The conference sought to provide more context for Edmontonians as the arena issue continued to move forward.

These events provided the backdrop for an ensuing discussion about the need for a new or renovated facility. Another issue woven into the debate related to the ownership of the team. In 1998, a groups of business leaders stepped up to buy the team from the previous owner, Peter Pocklington. The group took a stewardship approach to the franchise and as the notion of a new facility arose, questions were raised about the willingness and/or ability of all or some of the group to step up financially to fund a comprehensive arena development project. In the meantime, local drugstore magnate, Daryl Katz (Katz Group), made several attempts to purchase the team, which he eventually did in June, 2008. This may have also changed public perceptions of the role of the Oilers in the arena project, given that he was perceived to possess greater wealth and therefore the ability to fund the project privately.

As this chapter is written, the process for the arena is ongoing. The Katz Group has acquired or hold options on nearly twenty acres of downtown land, and is working to develop a design model which will presumably be brought to the city. The issue has received substantial coverage in the local media, with commentary and coverage appearing almost daily. It is also important to note the author's personal involvement in the arena issue in Edmonton. I was a consultant hired by the city of Edmonton in the fall of 2006 to provide an overview of arena development elsewhere, served on a subcommittee that helped to produce the *City Shaping* document, and also organized the conference held in February 2008. I have also made comments in the media regarding the arena issue.

4 Method

Theory on media framing undergirds the analysis of newspaper coverage of the arena development project in Edmonton, and borrows from the method established in Buist and Mason (in press). Framing has been employed across academic disciplines (Van Gorp, 2007), but is rooted in communication research (Bryant & Miron, 2004). Frames are "schemata of interpretation" that allow users to "locate, perceive, identify, and label a seemingly infinite number of concrete occurrences defined in its terms" (Goffman, 1974, p. 21), and can be defined as, "*organizing principles* that are socially *shared* and *persistent* over time, that work *symbolically* to meaningfully *structure* the social world" (Reese, 2001, p. 11). Thus, frames are found within texts and can provide context and create meaning for their audience (c.f. Gamson & Modigliani, 1989). They are "manifested by the presence or absence of certain keywords, stock phrases, stereotyped images, sources of information, and sentences that provide thematically reinforcing clusters of facts or judgments" (Entman, 1993, p. 52). Frames are seen in the manner through which issues are selected and presented to an audience, and the ways in which aspects of the issue are emphasized or ignored (Tankard, Hendrickson, Silberman, Bliss, & Ghanem, 1991; Gitlin, 1980).

The method used in the current paper employs a similar procedure to that used by Buist and Mason (in press). The City of Edmonton tracks media coverage of issues germane to the City, including that of the new arena project. Articles, letters to the editor, editorials, and other published commentary were obtained from the city of Edmonton in electronic format, representing 470 pages of coverage, from December 2008 through mid November, 2009. It is important to note that, while this represents a comprehensive database of all arena stories from this time period, it does not capture some of the earlier discussion in the local media. Thus, the results of this chapter should only be treated as a preliminary examination of the frames in Edmonton, as coverage remains ongoing as the issue remains a hot button topic in Edmonton. A modified content analysis (Suddaby & Greenwood, 2005) of both manifest and latent content was conducted. Berg (1998, p. 226) defined manifest content as, "the surface structure present in the message" and noted that latent content represents, "the deep structural meaning conveyed by the message". The focus on latent content is important as, "media texts often contain only portions of a frame and rely on audiences to infer the rest based on their existing cultural knowledge" (Edy & Meirick, 2007, p. 125; c.f. Gamson & Modigliani, 1987). Analysis was completed in several phases as outlined below.

Following Buist and Mason (in press), arena coverage from five local newspapers (the *Edmonton Journal, Edmonton Sun, Edmonton Examiner, See Magazine*, and *Sherwood Park News* were coded based on four frames identified by Buist and Mason (in press): (1) economic development, (2) civic status, (3) civic priorities, and (4) financing. The economic development frame typically includes discussion and commentary focusing on tangible economic benefits—such as employment opportunities, municipal tax revenue, and spin-off development—that would be generated by a proposed sports facility development project. Civic status frames discuss the role of the proposed facility in increasing the standing of the community as a whole. A civic priorities frame grounds debate within the broader context of community needs, such as other infrastructure requirements, and services such as education and health care. A financing frame considers the cost of the project, the appropriateness of using public funds to subsidize private industry, and the funding responsibilities of different levels of government. Thus, the first phase of data analysis involved a form of deductive analysis that looked for the presence of these frames.

In addition, given the current state of the global and regional economies, data were also coded for an additional frame—economic downturn. In the second phase of data analysis, an inductive approach was undertaken to determine if any additional frames existed in the Edmonton context. This occurred through the identification and classification of common words and phrases within the data collected (cf. Entman, 1993). It is important to note that a single data point (article, editorial, or letter) may have contained several frames. Finally, chronologically ordered matrices (Miles & Huberman, 1994) were constructed around the frames in order to organize the coding and analysis of data.

5 Results

The four frames from Buist and Mason are detailed below, along with findings from the economic downturn frame. In addition, two additional supporting frames, relating to the comprehensiveness of the project and the need for public consultation, are also discussed.

5.1 Economic Development

As in most arena debates, the issue of economic development figured prominently. However, in contrast to many other arena development scenarios, few proponents in Edmonton have relied on ambitious promises of economic benefits that would be obtained through the construction of the arena. Other research has shown that arena proponents have moved toward the use of intangible benefits to argue the merits of sports facilities in cities (Buist & Mason, in press). Thus, this may reflect a general trend in the discourse surrounding media coverage of facility development, and the influence of the academic researchers who have developed a body of literature that questions the economic impact studies commissioned by proponents.

Arguments for and against the downtown development tended to focus on the utility of an arena in revitalizing the downtown core, and whether or not the downtown needed revitalizing at all. According to the Mayor, "This is a project whose time has come .... I think it will be a remarkable thing for our city and downtown" (McKeen, 2008a, p. A10). Others suggested that they would be more likely to come to the city to watch events in a downtown facility (Sound Off!, 2008). Proponents, including the NHL's Commissioner, Gary Bettman, focused on the different groups that would benefit:

> A building that is used not just for hockey games but for concerts and family shows and conventions can revitalize an area in the city. It can attract tourists. And it is something that is really critical, I think, both for the future of the Oilers and the city (Jones, 2008, p. S3).

One city councilor argued that a downtown arena could be a catalyst for a much broader revitalization program (Landry, 2009), a sentiment quickly championed by Oilers President and CEO, Patrick LaForge: "From my point of view, and the project's point of view, it's about Edmonton and developing a part of a key area of our city in the downtown core .... It's an area that needs some refurbishing and redeveloping" (Van Diest, 2009, p. S4). Similarly, an unnamed source noted that:

> This is not just about a hockey team or an arena. The city's need to revitalize the core creates this once-in-a-lifetime opportunity, and a development like this only works if you have the kind of anchor tenant that an arena can provide (Lamphier & Kent, 2009, p. A1).

Thus, as the debate continued, the development was positioned by proponents as an opportunity for hockey fans and the team, along with those interested in attending other entertainment acts. It was also positioned as an opportunity to jumpstart other development in the downtown, including expansion of the local light rail transit system (Davies, 2009).

However, the economic development arguments put forth by proponents occasionally focused on more traditional arguments; for example, when Commissioner Gary Bettman gave a luncheon presentation sponsored by the Edmonton Chamber of Commerce in December of 2008. Bettman emphasized the need for a new arena, and the need for public dollars to subsidize it. Stepping away from the more conservative rhetoric used by local proponents, Bettman exclaimed "It is imperative that the Oilers have a new building .... It can become an economic engine. It can attract tourists. It is critical both for the future of the Oilers and the city" (Barnes, 2008, p. A1), adding that "I'm a big believer in downtown arenas. They are great economic energizers" (Jones, 2008, p. S3).

Even those in favor of the new development were wary of the display of exuberance, and raised questions about Bettman's role in the development (McKeen, 2008a):

> Edmontonians understand which foot the skate's on in this regard. Because of that, they will doubtless decide to back a privately financed replacement for shopworn old Rexall Place. The potential benefits for the whole community are not exaggerated by Mayor Stephen Mandel. But, please, leave us alone to make this major change to the fabric of our city on our own time, in our own way (Gary Bettman's free advice, 2008, p. A18).

Not everyone was enamored with the possibility of a new arena downtown, or agreed with the economic development benefits put forth by supporters. A focus of these criticisms was on the "dead time" created during non-event times and dates, and whether an arena was needed as an anchor for such a project (Griwkowski, 2009, p. 10). Others were quick to note that successful arena-anchored development projects were the exception, rather than the norm in other cities, and that while an arena development might increase the number of visitors:

> 'It's foot traffic,' says Brad Humphreys, a professor of economics at the U of A who lectures on the economics of sports arenas. 'But it's also drunks and a lot of people staggering around.'
>
> But Humphreys points to St. Louis and Washington, DC, as cautionary examples, places where major new baseball stadiums did little or nothing to spur promised residential projects. Both ballparks, he says, remain surrounded by huge vacant lots (Simons, 2009, p. A12).

Another interesting counterargument to the project suggested that it might be damaging in its own success; that is, it might successfully attract so much activity that it might harm other parts of the downtown, acting "as a vortex, sucking up investment and entertainment dollars, not to mention existing retail and office tenants, from the rest of downtown" (McKeen, 2009c, p. B1).

In sum, the economic development frame in Edmonton focused on the value of building the new arena downtown as part of an urban revitalization program. Both supporters and opponents have argued how they view an arena's role in generating a successful revitalization plan.

5.2 Civic Status

The civic status frame figured prominently in coverage of the arena issue in the local press. Edmonton has experienced several events in recent years that reveal an undercurrent of concern about the city's image to those outside of the city. One relates to a negative report about the city by a *London Daily Telegraph* reporter covering the IAAF World Track and Field Championships held in 2001. The article, titled "Deadmonton Comes Alive" described the Alberta capital as a "visually unappealing corner of Canada," igniting a firestorm of controversy locally. The article's author, Robert Philip, later wondered, "What puzzles this observer is that if Edmontonians are so confident about their city's greatness, why do they give a monkey's about what some visiting hack might think? But they do" (Reporter expresses regret, 2001, n. page). More recently, the highly publicized request of Oilers star player, Chris Pronger, who requested to be traded following the 2005–2006 season, has lead to concerns about the image of the city. This has been exacerbated by more recent attempts to sign and trade for high-profile players, who have accepted less money to play for other teams in other cities.

In terms of coverage of the arena issue in local newspapers, a focus has been on the progression of the city from a smaller, more familial community to an economically powerful, emerging city. Thus, debate surrounding the arena has been couched in terms of how the arena development can help the city reach a certain status amongst cities, and/or whether reaching this status is even something that matters to Edmontonians. For the most part, however, appeals to status were employed by those supporting the development project. As explained by one local columnist:

> Like all high-profile and expensive projects, the arena brings out the best and the worst in Edmontonians. The best: a willingness to debate. The worst: a tendency to disparage bold ideas and bold thinkers, to wallow in negativity and self-loathing, to return to the abiding, operatic theme of potholes (Babiak, 2008, p. B1).

For proponents, site selection was frequently articulated in terms of vibrancy, where supporters such as Mayor Stephen Mandel noted that

> Edmontonians... sometimes forget how important a vibrant downtown is to a city's reputation and sense of itself (McKeen, 2008a, p. A10).

In this manner, opponents were branded as hicks who could not contemplate the vibrant world-class city that an arena development project could build toward, and other cities with busier downtown cores were cited as examples of what Edmonton should aspire to (McKeen, 2008a). As a result, the cost of not supporting the project was the (perpetuation of the) city's poor image elsewhere.

The civic status frame was also coupled with the financing frame, as many opponents were quick to note that the many of the cities that Edmonton aspired to—such as Vancouver—had built their own facilities using private funds (Jones, 2008). One author also tapped into Edmonton's rivalry with Calgary in articles appearing in both Edmonton and Calgary papers, noting that "the shovel isn't in the ground yet—but the fact the Oilers now have earth to move is enough to have Calgary fans drooling with envy" (Platt, 2009b, p. 2), and that "Surely Calgarians can covet just a little, as Edmonton moves a huge step closer to replacing Rexall Place with an ice palace that will make the Saddledome look like the wooden skate shack at the local outdoor shinny rink" (Platt, 2009a, p. 23). In this manner, the arena project was articulated in terms of delivering a certain "world" or "first-class" status to Edmonton (Klein, 2008), along with "all kinds of social, cultural and economic benefits to Edmonton" (Lamphier & Kent, 2009, p. A1).

The notion of changes to the city's position within a hierarchy of cities also figured prominently in coverage of the debate, where terms like world class

(Lamphier, 2009b) figured prominently: "'People like to throw around the word world-class a lot, but I think (the arena project) should actually be measurable to that standard and not just Edmonton-class, or what we've experienced in the past,' says urban planner Dave Onishenko" (Staples, 2009a, p. E4). This rhetoric was widely used by Oilers CEO Patrick LaForge, who suggested that a comprehensive arena development would make Edmonton the "Greatest Northern City" (Kent, 2009), making the city much more attractive for tourism, relocation and investment (Van Diest, 2009; Staples, 2009a). To emphasize the need for a new facility, other Oilers executives touched on the notion of status to express the franchise's needs: "'We're a world-class city now and unfortunately our building is not world class,' Oilers hockey operations president Kevin Lowe has said" (Staples, 2009c, p. E1). Similarly, Oilers owner Daryl Katz emphasized the broader development, claiming that: "Like a lot of Edmontonians, we are very excited about the prospect of a revitalized downtown and a new multi-use development that can make Edmonton the place to be" (Jones, 2009, p. S3). Even players spoke to how the development could elevate the city and the downtown: "'I'm an Edmontonian and I think it's a great thing for the city,' said [Oilers defenseman] Jason Strudwick.... 'It will continue the revitalization of downtown that we have to keep pushing. We want to have a good, vibrant downtown'" (Tychkowski, 2009, p. S3).

Status arguments were not limited to local politicians and those associated with the franchise itself; for example, one letter to the editor (from "Frank") appealed to Edmonton's inferiority complex and tied this directly to the downtown:

> I fully support a new downtown arena and would be more than happy if my tax dollars went into funding it as opposed to things I never use, i.e., ETS, homeless shelters, public schools, etc. I realize that those things are probably more important than the arena but I pay for them through my tax dollars and get no benefit. People do not seem to realize that our downtown is the single biggest reason that our city has an image problem. I've been to every major Canadian city and Edmonton's downtown is truly embarrassing when compared to Toronto, Montreal, Vancouver, and even Calgary. The downtown arena will not solve this problem; however, it is a giant step in the right direction (Sound Off!, 2008, p. B3).

Another letter claimed that "Edmonton is on the cusp of making the transition from a mid-sized prairie city to a dynamic major Canadian city. The new arena will mean a new future in Edmonton" (Davies, 2009, p. A15). Similarly, another wrote that "The proposed new arena and entertainment district has the opportunity to create a dynamic and animated public place that will become a major focal point for our downtown and the region" (Dulaba, 2009, p. A16).

Some of those who saw the arena development as an opportunity were also concerned about the lack of architectural distinctiveness of the city, a concern that could be addressed by a spectacular arena and district design (Staples, 2009a) that could overcome Edmonton's bland, anonymous streetscapes (Towering Vision, 2009). Others cautioned that it was smaller cities that felt a sense of inferiority that tended to subsidize these developments (Staples, 2009a). Even those not overtly supporting the project acknowledged that the arena development provided an opportunity "to build an architecturally striking, world-calibre sports and concert facility, and a chance to rejuvenate our city core" (Simons, 2009, p. A12).

The debate over control and operations of the building also became bound up in notions of world-class status and the caliber of facility and the events held there. While Northlands continued to operate the existing facility and anticipated having some participation in the new one, grumblings out of the Oilers organization hinted that the new facility would be out of the non-profit's league: "The implicit message: Northlands may run a decent show in Edmonton, where it's insulated by government largesse, but if the city's new arena complex is going to be truly world class, it needs to be run by a world class [arena management] team" (Lamphier, 2009b, p. A1). This underlying message was then clearly articulated by Oilers CEO, LaForge:

> We're interested in being the best in the world. I'm not trying to poke Northlands in the eye here. ... Northlands has done a good job so far. Nobody would have an issue with that. (But) the new world is big operators like AEG, Live Nation and the like. They are creating a new world of preferred outlets that they bring their best Triple-A entertainment to, and we want our place to be one of those stops (Staples & Lamphier, 2009, p. A1).

Thus, notions of world class and the status of the city were used as a way of articulating the benefits that the new development would provide, while also providing a means for stakeholders to engage one another around their possible roles in the project.

5.3 Civic Priorities

The civic priorities frame was consistent with that found in other arena development debates, where justifications for the use of public funds were couched in terms of their opportunity costs (see Buist and Mason, in press). For opponents in Edmonton, this seemed to put the arena project at odds with basic services, and also became bound up in general dissatisfaction with the performance of local politicians. As the city moved into the economic recession, some columnists voiced concerns over spending approved by city councilors: "A majority of them agreed to borrow $1 billion for civic projects—almost half of it for palatial new recreation centres. In doing so, council established a new financial benchmark in the public consciousness" (McKeen, 2008a, p. A10). Because the city agreed to push forward with several large-scale infrastructure development projects in spite of economic uncertainty, the likelihood of putting additional funds towards an new downtown arena seemed more remote. As suggested by Councilor Linda Sloan, "I don't see that kind of facility serving the needs of Edmonton. ... If the Oilers or if the National Hockey League want to contemplate building a facility, I think it's up to that sector to build it" (Landry, 2008, p. 2). Councilors from neighboring communities agreed; Jason Gariepy of Sherwood Park cautioned against nearby Strathcona County being brought into the funding issue (Di Massa, 2009). As explained by Gariepy himself, "I don't believe that City of Edmonton taxpayers are willing to service a big fat loan for the arena. I don't believe there is an appetite to use public funding for a private venture" (Gariepy, 2009, p. 9).

With funds becoming more scarce, other local politicians questioned the use of public funds to subsidize professional sports franchises (Diotte, 2009a). Locals also wrote into the newspaper to vent about the potential use of funds for an arena when they could clearly identify other, more pressing needs: “Instead of spending hundreds of millions of dollars on a hockey rink we don’t need, how about spending that money on a couple of bridges that would make it easier to get into and out of downtown?” (Venting, 2009d, p. B12). Similarly another opined that “Instead of spending hundreds of millions of dollars on a hockey rink we don’t need, why not spend that money on improving the Yellowhead [Highway] instead? The traffic is getting worse every day, with no improvement in sight” (Venting, 2009b, p. B2). Others felt that the arena project would serve too few to justify any public investment: “An arena serves relatively few people and should not be the centre of the city. Not only is this a poor use of the property, but taxpayers should hold the mayor and council more accountable for these ridiculous, self-serving projects” (Dubeta, 2009, p. A17). Building on the sentiment that the arena project served too narrow a constituency, others complained that putting an arena downtown would cater simply to “rowdy fans”, whereas “A diverse atmosphere of people and activities in the core would appeal to a more diverse population of visitors” (Stewart, 2009, p. A16). Concerns regarding the gentrification of neighborhoods also arose, with suggestions that “it would take serious political will to enforce any social housing or similar conditions on this kind of proposal” (Warnica, 2009, p. B1).

The fact that the team and its players would benefit heavily from a new facility was not lost on many observers:

> How often do we allow someone to reach into our pockets and pull out not only our loose change but our credit card as well, so we can spend tomorrow’s dollars to help fund a small number of multimillionaires who try to convince us that it is necessary if we want to maintain our privileged position of having major league sports franchises in our city? (Fan Forum, 2008, p. C2).

In addition, some expressed an underlying resentment towards the Oilers organization in general, where the team and its management were chided for arbitrarily deciding what was best for the local community (including a new arena downtown). One comment went so far as to accuse the Oilers, who created a charity lottery, of hurting another local charitable lottery by saturating the market for lotteries in the city (Fan Forum, 2008).

The Oilers were more careful to try to disassociate from other controversial civic debates, including the issue of closing the city’s downtown municipal airport. When Oilers Chairman, Cal Nichols, emerged as a champion for keeping the airport open, concerns were raised about how this would be viewed in light of the team’s desire for a public subsidy for the arena (McKeen, 2009a; Staples & Gerein, 2009). As a result, Nichols relinquished his duties with the Oilers, acknowledging that “There’s a number of people on the political side in City Hall that are of opposing view to me [in keeping the airport open], but at the same time (the) Oilers and Katz Group are trying to work with them on the development of a new downtown arena district and my being around the table is becoming an issue” (Virag, 2009, p. 3).

To summarize, the arena debate was framed in terms of civic priorities as proponents prioritized the revitalization of the downtown; in their view, those unable to view the potential for the arena project to do so simply could not "see it yet" (Mackinnon, 2009a). In addition, opponents were considered cranks who opposed all kinds of forward-thinking development:

> Bidding on Expo 2017 is too much money, bidding on the Universiade was too costly, and a hockey arena downtown is too expensive and will cause traffic problems. The LRT [light rail] is too futuristic, the Henday [ring road] is too noisy, and the cost of the 23rd Avenue interchange is prohibitive. Finally, the City Centre Airport is the jewel of Edmonton. Do you think maybe Tourism Calgary hired all these people to keep us stuck in the 1970s? (Venting, 2009a, p. B2).

In contrast, opponents clearly outlined the infrastructure and groups that would be better serviced by the funds that would need to go into the project, which would be primarily for the benefit of the team and its players. As the economy continued to decline, this only exacerbated the problem for some critics (to be discussed in greater detail below):

> The front page of Saturday's *Journal* summed up the obscenity of financial management in the province of Alberta. The main headline is about the Katz Group and Northlands contending for the $1-billion prize of managing the new $450-million downtown arena; an arena that will largely be funded by taxpayers. A story lower on the page says Education Minister Dave Hancock may insist on cuts deeper than the $44 million already sought, with rumours indicating cuts will actually total hundreds of millions of dollars (Griwkowski, 2009, p. 10).

5.4 Financing

By far the most dominant frame in Edmonton newspaper coverage related to financing—who would be expected to pay for the arena development project and how the financing model would operate. Given that there is no such model in place at the time of this writing, all of this discourse remains speculation. From the perspective of arena supporters, the use of public money was justified, as there had been a history of public funding for sports facilities in the province, and comparisons were drawn to the new, publicly funded art gallery (Staples, 2009b). The team argued that it needed more arena-related revenues to reduce the disadvantages it had in comparison to other league clubs (Staples & Lamphier, 2009). This was reinforced by NHL Commissioner, Gary Bettman, who proclaimed that "There is no way a building can be built here without a significant public investment" (Jones, 2008, p. S3). Pro-arena supporters responded to opponents by arguing that tax revenues already went to many services that that citizens did not necessarily use:

> Let's get it done already! To all those who don't want their tax dollars funding it I say, well, a lot of my tax dollars go to fund all sorts of stuff I don't and will never use and don't want to fund yet my tax dollars still go there so that others in the community can enjoy these things. That's life. Let's just do it already! (Sound Off!, 2008, p. B3).

However, there was a strong response from those who did not want to see any tax dollars going towards the project, especially where the beneficiaries were private entities: "as nice a fellow as Mr. Katz may be, he doesn't deserve a free boost of hundreds of millions of dollars to help further milk the population behind closed doors" (Griwkowski, 2009, p. 10). This argument was not limited to non-hockey fans; in many cases those opposed were strong supporters of the team: " I love the Oilers but I strongly object to our tax dollars building a play-ground for millionaire players and a billionaire owner. There are much better uses for our money" (Venting, 2009c, p. B2). This suggested that opposition might be linked to resentment fans had towards the salaries paid to players. Meanwhile, the core debate remained the appropriateness of using taxes to pay for the facility (Barnes, 2008).

Other groups, such as the Canadian Taxpayers Federation, became more vocal as the debate ensued; Alberta Director, Scott Hennig, professed that "Our only concern is that taxpayers' money not go into building a new arena," and he cited other recent privately funded NHL arena development in Canada (Staples, 2009a, p. E4). In response, supporters tapped into the city's alleged inferiority complex and civic status as a way of swaying opinion on the use of public funding:

> If public money is deemed to be necessary, though, winning the hearts and minds campaigns will be crucial, Onishenko says, and the battle won't be easy, given the negativity in Edmonton. "We are a city of disbelievers and self-doubters at times, who like the growing vibrancy of our southern counterparts, as well as Vancouver, Toronto, but don't believe we can achieve something similar here, and are content to settle with our urban monuments of mediocrity" (Staples, 2009a, p. E4).

The debate also shifted from the use of taxes in general, to the *types* of taxes required or to be used to service any public debt.

For the most part, letters to the editor were not concerned about the form of the tax; rather, they viewed the anticipated need for public money as directly affecting their finances:

> As I'm driving down the heavily constructed 111th Avenue, trying desperately to figure out if any of the side streets are open for traffic, I'm contemplating this latest development scheme for the downtown core. Not being a mathematical genius, I'll use simplistic math because that seems to work for everyone else. If the new development costs $1 billion and the new owner chips in $100 million, that leaves $900 million owing. It's easy math–$900 million, three million Albertans, a mere extra $300 for every man, woman, child. In my household of three, that would only be $900. Collection would be the city's problem (Pryer, 2009, p. A18).

Others were more succinct: "Let the business suits play at development and leave me and my taxes out of the equation. I don't trust the mayor when he says 'no current taxes' will be used for the arena" (Taylor, 2009, p. A18). Another commented that "While the arena may be spectacular, citizens' awe will end when traffic problems become a daily frustration, and when they are called upon to pay for the arena through their taxes so a few people can get richer" (Kisilevich, 2009, p. A19). Concern over tax increases was also linked directly to a recent general property tax increase (Gerard, 2009; Hanon, 2008).

The backlash was exacerbated when advice was given from others on the issue. This was directly particularly harshly towards the NHL Commissioner. When

Bettman made his announcement re: the need for public funding, one response was "Let Bettman move to Edmonton and pay his taxes toward that stupid arena" (Venting, 2008, p. B2). As another wrote:

> Wow! Thank you Mr. Bettman for your wonderful insight on our city. Maybe on your way, you may have noticed we need other things more important than another building for millionaires. Maybe they can put their own $$ into the building and leave us taxpayers alone. You know the taxpayers that cannot afford the luxury of watching grown men slap a piece of rubber around and get paid millions for it. You know, ordinary people that have nothing to do with this building, and only want to support their own families. Maybe another homeless shelter would be better suited. I know it would certainly be more needed (Sound Off!, 2008, p. B3).

Thus, other frames such as civic priorities were employed in support of the finance frame. In the same coverage, another commenter suggested that it was "Easy to say, Mr. Bettman, when you won't be affected in any way, shape or form financially if this fiasco goes ahead. We will be the ones choking down another insane tax hike—mark my words" (Sound Off!, 2008, p. B3).

As the issue of funding continued to swirl, the Mayor became outspoken about the use of taxes, reporting that the new development would have no impact on property taxes (McKeen, 2008a). While the use of a community revitalization levy had been suggested in March of 2008 with the release of the *City Shaping* report, opinions remained split on the utility of such a mechanism:

> No, your property taxes won't rise. The only taxes supporting the project will come from the project itself. No project, no taxes to fund it. I'm fine with this idea. But many Edmontonians are not. More important, a number of city councillors are seriously leery (McKeen, 2008b, p. A11).

This was viewed by others as a form of creative accounting. Thus, although Mandel clearly stated that "If you live in Edmonton, your current property tax dollars will not go into it, nor will mine" (Landry, 2008, p. 2), some were cautious:

> Pardon me, Mayor Mandel, if I don't applaud. You promise us that none of the arena's estimated $450-million price tag (not including at least six acres of prime real estate) will be paid for with what you term "current property tax dollars." You'll have to forgive us for cringing at the word "current." (Hanon, 2008, p. 10).

The proposed community revitalization levy (CRL) would use incremental increases in tax revenues from the development surrounding the proposed arena to service public debt. Although still a tax, this would reduce the direct financial pinch taxpayers would feel. Proponents argued that it allowed those who supported the project—by living, shopping, attending events—to bear the public's burden for the development. This also served to create the need for a much larger development project, as there would need to be enough alternative types of development to generate the necessary taxes to service the debt (McKeen, 2009c). However, for the duration of the newspaper coverage reviewed in this chapter, no definitive model was developed that explained how the CRL would work. Instead, the Mayor invited speculation with comments like: "There's a variety of methods to do it. I think at this point in time it's premature to discuss how it can be funded. We'll leave it as it is" (Landry, 2008, p. 2). As a result, the proposed project was described as being funded with "creative government cash" (Hicks, 2009b, p. 6).

The lack of a formal funding model and uncertainty about the use of a CRL versus other forms of public funding created a demand for more transparency and a lack of trust on the part of stakeholders in the community. Columnists began to question the benefits to the city, and the motives of the team owner, especially with no large scale vision that could be used to develop the CRL: "That sounds good, in theory. But where's the guarantee from the developer (i. e., Katz Group) that there would be any upfront spinoff projects that would be built alongside the new arena?" (Lamphier, 2009d, p. E1). Others became more critical of the wording that Mandel was using when promoting the arena:

> Go all the way back to Omniplex [another arena development project], defeated in a referendum, and you find a citizenry leery of seeing its tax dollars used on big public projects. Mayor Stephen Mandel promises that your taxes won't fund the arena project. But let's be clear: Taxes will be used to fund the project (McKeen, 2009b, p. B1).

Another suggested that:

> Given the exorbitant tax increase just foisted on us, along with the uncertainty of the economy, surely you'll understand our skepticism over assurances that your new arena won't dig into our wallets. When you finally unveil your financing plan, the question taxpayers will ask is: what's really in it for us? (Hanon, 2008, p. 10).

As the issue dragged on, more speculation continued and a general impatience developed regarding knowing what the city would ultimately have to pay into the project and what the components of the development would be (Lamphier, 2009a; 2009d), a process described as the "great riddle of the Edmonton universe" (Hicks, 2009c, p. 6). To this point, speculation has lead to a lack of trust for the parties most directly involved in the project (Diotte, 2009b). For this reason, some questioned whether the project would be able to receive widespread public support (Fekete, 2009).

In terms of political support, while the Mayor clearly was a champion of the proposed development, others in the city and at other levels of government were less enthusiastic. In fact, Alberta Premier, Ed Stelmach, directly addressed the issue, stating that "It's very clear we're not putting money into arenas" (Staples, 2009b, p. A1). Even political candidates from other parties attempted to distance themselves from the use of funds for the arena project (Diotte, 2009c). In October, 2009, the Mayor of Quebec City announced his intentions of building a new arena, funded by several levels of government. This did not alter the position taken by the Premier in Alberta: "'The request by the Quebec City mayor has no impact on the premier's position,' says Tom Olsen, Stelmach's media relations director. 'His position was and continues to be that there will be no direct public financing of privately run hockey arenas in Alberta'" (Lamphier, 2009f, p. E1). Given that Mayor Mandel also promised that "We would not use grant money from the federal government (provided) for other projects" (Lamphier & Kent, 2009, p. A1) it is understandable that speculation regarding the burden faced by local taxpayers would grow.

Another issue related to financing that arose was concerned with the involvement of Northlands. As operators of the existing facility, some observers felt that their involvement was necessary to make to project work financially. Councilor Ed Gibbons, a Northlands board member, doubted that city council would invest in the project without Northlands' involvement (Lamphier & Kent, 2009). Andrew Huntley, Chairman of Northlands, also argued that his organization's involvement would help to make the project more viable, and aid in acquiring the financial support of higher levels of government (Staples, 2009e). This discussion also related to the question of which party would own the facility: the city, the team, the arena management company, or a development corporation (Kent, 2009).

Other coverage discussed possible new revenue streams, such as personal seat licenses (Hicks, 2009a) and the use of casino revenues to service the debt. As the issue moved forward, speculation arose regarding the size of the development and the ability of increasing tax revenues to service debt (Lamphier, 2009c; Staples, 2009d). However, through late 2009 no concrete financial model has emerged. As explained by on Katz spokesperson:

> Funding models have been discussed, studied and pursued, but that remains very much a work in progress and could go any number of ways. There have been extensive discussions with Northlands, and we have also talked with the city and province, about roles, funding options, etc., but nothing definitive has come of those talks (Lamphier, 2009e, p. E1).

5.5 Economic Downturn

As explained above, in addition to coding the data for the four stadium frames identified by Buist and Mason (in press), newspaper coverage was also examined to determine the degree to which the broader economic environment was also influencing the newspaper discourse surrounding the arena development in Edmonton. The economic downturn influenced the frames described above, particularly the civic priorities frame. The region had undergone a substantive boom in the years leading up to 2008, driven by development in the oil sands in northern Alberta. As a result, the global economic crisis had not hit the city and province as devastatingly as other parts of the country. In March of 2008, provincial Finance Minister, Iris Evans, claimed that "Alberta will be relatively well-off because we're safer. ... I think the long-term investments in the oilsands really protect us, so there's a sustainability to our economy that isn't there elsewhere" (Audette, 2008, p. A3). However, by late 2008 the crisis had clearly affected the local economy. One issue that was raised was how, during the boom, the city had not taken advantage of the opportunity to develop more arts, cultural, and entertainment amenities. One local columnist commented that "It was painful to come through a boom without anything to show for it but an art gallery" (Babiak, 2009, p. B1).

However for some, the recession increased the need for a new arena development project, which could be justified on the grounds of cheaper construction costs, which would lower the costs of the overall development (Fekete, 2009). This logic was shared by the Mayor, who claimed that the new arena would create jobs in a tough economic climate (McKeen, 2008a). After arriving in Edmonton to give a speech at a local luncheon, NHL Commissioner, Gary Bettman, suggested that Edmonton had weathered the recession well and was poised to build a new arena: "You build buildings for the next 30 years," Bettman told the press conference. "If you think you're going to be looking at a 30-year depression you certainly don't do anything" (Jones, 2008, p. S3). According to some reports, this sentiment was shared by some on city council (McKeen, 2008a).

However, to others the logic of spending $400 million on a new arena during an uncertain time, even if the region had not been hit as hard as others, was difficult to follow. To them, building such a significant project was unnecessary, especially with several major oil sands development projects stalled (Hanon, 2008). In addition, concerns were raised that there would not be enough discretionary money available to both fans and corporations to sustain the new facility. As one fan wrote:

> Bettman, Daryl Katz, Pat LaForge and Co. should be embarrassed to even suggest the construction of new arena for Edmonton at this time. Do they really think businesses and taxpayers would contribute even greater amounts than they are now for a facility that would cost users even more money to access? (Fan Forum, 2008, p. C2).

The economic downturn also increased public scrutiny on public expenditures of all kinds, including capital grants that both NHL teams in Alberta had received through facility renovations (Diotte, 2009a). While many continued to acknowledge that the facility would require some form of public subsidy, more argued that the subsidy should be borne by hockey fans and not the taxpayers at large (Sound Off!, 2008). As the recession continued, letters to local newspapers questioned the value in pursuing such projects given the need to cut back other services:

Faced with a budget deficit, it is proper for city council and administration to try to cut expenditures and increase revenues with the least negative impact on residents. While I have concerns about some of the specific suggestions that have been put forward, my real beef is more general. That the city is going through this exercise, while continuing to pursue grandiose schemes (the Expo bid and pursuit of a downtown arena, to name but two) which will cost Edmonton ratepayers undisclosed millions, is ridiculous, if not outright negligent (Letters to the Editor Column, 2009, p. 10).

The issue was exacerbated when it was articulated as a potential tax increase to "fund a new playpen for rich hockey players" (Lamphier, 2009a, p. E1).

6 Emerging Frames

While the four frames identified by Buist and Mason (in press) and economic downturn frame figured prominently in the newspaper discourse surrounding a new arena in Edmonton, two other frames emerged from the analysis of the newspapers during the time period covered. The first related to the comprehensiveness of the total urban infrastructure development. This was a key for some who, while not in support of a stand-alone facility, discussed a much larger, integrated development and its implications for economic development and civic status. For members of the Katz Group more broadly, and the Oilers more specifically, this was seen as a means to garner wider support from those not interested in the local hockey team:

> 'This is very much about the city,' said Josh Pekarsky, spokesperson for the Katz Group, on Wednesday. 'It's about far more than an arena. It's about an entertainment and sports district for the city and for the region. There are a number of options being considered.' (Babiak, 2008, p. B1).

In addition, supporters of the arena were then able to articulate its importance in terms of integrating with planned or existing downtown development projects, such as the new art gallery (McKeen, 2008a), and to promote an arena district as a means of revitalizing the downtown on a large scale (Hicks, 2009b; Lamphier & Kent, 2009). In addition, the arena district became embedded in a broader discourse of other Edmonton concerns, including sprawl and the need for greater density (Lamphier & Kent, 2009), and included, at various times, discussion of hotels, performing arts theatres, student housing, a casino, a hockey centre of excellence, hotels, and convention centre development (Jones, 2009, Lamphier, 2009b).

Comments from outside experts thus focused on the need for the city to develop a comprehensive development plan with "the right mix of amenities, such as parks, condominiums, restaurants and stores, creating a city neighborhood where people will want to live and to visit" (Staples, 2009c, p. E1).

A final frame that emerged from the analysis was broadly concerned with public input; that is, regardless of whether or not one supported the proposed project or not, there was a building consensus that it should ultimately be the taxpayers that should decide if any public funds should be used (Loome, 2009). This was put forth more so by opponents, and was often couched in terms of the use of funds to pay the salaries of millionaire hockey players and the team's billionaire owner.

Another reason why more public input was sought related to trust—trust that the Oilers ownership was really looking after the interests of the public in proposing such a large development project, and also related to the faith that locals had in the political leadership to make the right decision:

> So why isn't the population jumping on board enthusiastically? One word: trust. Edmontonians don't trust that they won't get dinged for a tax hike over the development. They don't trust the word of Mayor Stephen Mandel on the issue although he's constantly promised no tax money will go into the thing. Nor do they trust Katz who's almost a Howard Hughes-style recluse. Sadly, history teaches that it's not always wise to trust politicians or sports team owners (Diotte, 2009b, p. 14).

Similarly, the larger the scale of the project proposed, the more input sought:

> The arena's proponents would like to present their bold plan as a fait accompli. But this project will have a radical impact on the shape of our downtown. No matter who pays for the arena, or who profits from it, we citizens should–and do–have the right to decide if this is indeed an appropriate site for this urban megaproject. Just because this particular property deal suits the financial interests of those who are buying the land and those who are selling it doesn't make it good urban planning, or good for the rest of us (Simons, 2009, p. A12).

This view was supported by the Mayor, a vocal proponent. In an interview on a local radio station, Mayor Mandel clearly stated that the city would have a "big time say" in the development if public funds were to be used (Staples, 2009d). In addition, the secretive manner in which the team's leaders sought to explore the project also served to draw criticism from those seeking greater public input. As one columnist explained:

> Somehow, a project with the potential to transform one's city should be pitched to the citizens not merely by the billionaire who has promised to kick-start it with $100 million.
>
> This isn't merely a matter of protocol. It has to do with buy-in by the population, and that's critical here.... That's fabulous, but who elected the Oilers, Katz or LaForge and gave them the mandate to make downtown more livable? (Mackinnon, 2009b, p. C1).

7 Discussion and Conclusions

The examination of newspaper coverage in Edmonton over the past year has revealed that many of the same frames found in other debates over the public subsidization of sports facilities in other cities are present in the current context. In the case of Edmonton, because the process is ongoing and has been drawn out over time, additional speculation regarding the financing of the project has become a dominant frame. Interestingly, bold claims of economic impacts have not been used by proponents to the same degree as elsewhere; however, as the project moves forward this does not preclude some of these claims from being made.

Despite the popularity of hockey and the Oilers in Edmonton, there does not appear to be an appetite for the use of public funds to pay for a new arena development, nor does there appear to be a sophisticated understanding of how the funding mechanism will work. This is due to the lack of a definitive model being introduced to this point, but also a reflection of a general lack of support for public funding of any kind. While this has been exacerbated by the changes to the broader economy, there is little evidence that there was widespread support for such funding when the economy in Edmonton was strong.

As can be found in newspaper discourse of arena development projects in other cities, coverage of the arena issue has been more generally positive in local coverage; this can be partially attributed to the local newspaper's role in the broader urban development agenda. However, many letters to the editor remain unabashedly against the development. As the project moves forward, it appears that proponents will have to focus on transparency and building trust with the community in order to be successful. It also appears that those directly involved in the project will have to actively seek out the input of the community to make the development more in the interests of Edmontonians (Mackinnon, 2009a). It is important to note that this chapter has only focused on a portion of the coverage of an ongoing issue. If and when the issue is resolved in Edmonton, a content analysis of the entire body of newspaper coverage might yield more insights into the balance found of media coverage.

There were several themes that were layered through the frames. One relates to the use of public funds to pay (overpaid) hockey players. Although the team's players only indirectly benefit financially from a new facility, resentment toward millionaire players remained an undercurrent in negative reaction to the project. It will be interesting to see how the frames evolve in Edmonton as development plans become more concrete and the city and economy emerge from the global recession.

References

Altshuler, A., & Luberoff, A. (2003). *Mega-projects: The changing politics of urban public investment*. Washington, DC: Brookings Institution Press.

Audette, T. (2008, December 31). The Top News Stories 2008; Upsets in politics nearly matched upsets in global economics. A look back on the year that was. *Edmonton Journal,* A3.

Baade, R. A. (2003). Evaluating subsidies for professional sports in the United States and Europe: A public-sector primer. *Oxford Review of Economic Policy, 19*, 585–597.

Baade, R., & Dye, R. (1988). An analysis of the economic rationale for public subsidies of sports stadiums. *Annals of Regional Science, 22*, 37–47.

Baade, R., & Dye, R. (1990). The impact of stadiums and professional sports on metropolitan area development. *Growth and Change, 2*, 1–14.

Babiak, T. (2008, December 11). The Quarters concept an ideal arena to debate our future; A bold vision is vital to erase the architectural blight in the heart of Edmonton, whether it includes a hockey arena or not. *Edmonton Journal*, p. B1.

Babiak, T. (2009, June 9). Stalled Quarters project needs two things to get it going; Closing City Centre Airport and putting Expo 2017 site on Riverside golf course would do the trick, says Qualico exec. *Edmonton Journal*, p. B1.

Baim, D. V. (1994). *The sports stadium as a municipal investment.* Westport, CT: Greenwood Press.

Barnes, D. (2008, December 17). New arena for Oilers 'imperative,' Bettman says; NHL boss puts his mouth where his money isn't. *Edmonton Journal,* A1.

Berg, B. L. (1998). *Qualitative research methods in the social sciences.* Toronto: Allyn and Bacon.

Bryant, J., & Miron, D. (2004). Theory and research in mass communication. *Journal of Communication, 54*, 662–704.

Buist, A. N., & Mason, D. S. (in press). Newspaper framing and stadium subsidization. *American Behavioral Scientist.*

Coates, D., & Humphreys, B. R. (2000). The stadium gambit and local economic development. *Regulation, 23*(2), 15–20.

Coates, D., & Humphreys, B. R. (2003). *Professional sports facilities, franchises, and urban economic development* (UMBC Economics Working Paper 03-103).

Colclough, W. G., Daellenbach, L. A., & Sherony, K. R. (1994). Estimating the economic impact of a minor league baseball stadium. *Managerial and Decision Economics, 15*, 497–502.

Danielson, M. N. (1997). *Home team: Professional sports and the American metropolis.* Princeton, NJ: Princeton University Press.

Davies, R. (2009, September 1). Downtown arena has big potential. *Edmonton Journal*, p. A15.

Delaney, K. J., & Eckstein, R. (2003). The devil is in the details: Neutralizing critical studies of publicly subsidized stadiums. *Critical Sociology, 29*(2), 189–210.

Delaney, K. J., & Eckstein, R. (2008). Local media coverage of sports stadium initiatives. *Journal of Sport and Social Issues, 32*(1), 72–93.

Di Massa, M. (2009, October 16). Strathcona won't pay for new arena. *Sherwood Park News*, p. 1.

Diotte, K. (2009a, February 26). Tory spending slammed. *Edmonton Sun*, p. 7.

Diotte, K. (2009b, September 4). Don't keep us in the dark. *Edmonton Sun*, p. 14.

Diotte, K. (2009c, September 27). Alliance candidates score with arena tax ban. *Edmonton Sun*, p. 5.

Dubeta, R. (2009, July 14). Silly hall. *Edmonton Journal*, p. A17.

Dulaba, C. (2009, September 30). Traffic shouldn't be a problem at downtown arena. *Edmonton Journal*, p. A16.
Duquette, G. H., & Mason, D. S. (2008). Urban regimes and sport in North American cities: Seeking status through franchises, events, and facilities. *International Journal of Sport Management and Marketing, 3*(3), 221–241.
Edy, J., & Meirick, P. (2007). Wanted, dead or alive: Media frames, frame adoption, and support for the war in Afghanistan. *Journal of Communication, 57*, 119–141.
Entman, R. M. (1993). Framing: Toward clarification of a fractured paradigm. *Journal of Communication, 43*, 51–58.
Euchner, C. C. (1993). *Playing the field: Why sports teams move and cities fight to keep them.* Baltimore: The Johns Hopkins University Press.
Fainstein, S., & Stokes, R. (1998). Space for play: The impacts of entertainment development on New York City. *Economic Development Quarterly, 12*, 150–165.
Fan forum (2008, December 29). *Edmonton Journal*, p. C2.
Fekete, J. (2009, April 12). Calgary, Edmonton dealing with rink-construction dilemma; Arenas among smallest, oldest in league. *Edmonton Journal*, p. C3.
Friedman, M. T., & Mason, D. S. (2004). A stakeholder approach to analyzing economic development decision making: Public subsidies for professional sport facilities. *Economic Development Quarterly, 18*, 236–254.
Friedman, M. T., & Mason, D. S. (2005). Stakeholder management and the public subsidization of Nashville's coliseum. *Journal of Urban Affairs, 27*, 93–118.
Gamrat, F. (2002, July). Private financing for a new Penguins arena. *Allegheny Institute Report#02*-08. Allegheny Institute for Public Policy.
Gamson, W. A., & Modigliani, A. (1989). Media discourse and public opinion on nuclear power: A constructionist approach. *The American Journal of Sociology, 95*, 1–37.
Gariepy, J. (2009, October 16). Paying for new arena? *Sherwood Park News*, p. 9.
Gary Bettman's free advice. (2008, December 18). *Edmonton Journal*, p. A18.
Gerard, R. (2009, January 4). Looking for a legacy. *Edmonton Journal*, p. A13.
Gitlin, T. (1980). *The whole world is watching: Mass media and the making & unmaking of the new left*. Berkley, CA: University of California Press.
Goffman, E. (1974). *Frame analysis: An essay on the organization of experience*. New York: Harper & Row.
Gratton, C., & Taylor, P. (2000). *Economics of $ports and recreation*. London: Taylor and Francis.
Griwkowski, F. (2009, September 10). Burying socialized sporting facilities; Building the downtown arena underground would provide a beautiful public space above. *See Magazine*, p. 10.
Hannigan, J. (1998). *Fantasy city: Pleasure and profit in the postmodern metropolis*. New York: Routledge.
Hanon, A. (2008, December 16). Ice this idea before it gets out of hand. *Edmonton Sun*, p. 10.
Harvey, D. (1989). From managerialism to entrepreneurialism: The transformation in urban governance in late capitalism. *Geografiska Annaler, 71B*, 3–17.
Hicks, G. (2008, December 17). New arena great if someone else pays. *Edmonton Sun*, p. 6.
Hicks, G. (2009a, September 1). Patience is key to a worthwhile arena. *Edmonton Sun*, p. 6.
Hicks, G. (2009b, August 27). Katz talking two arenas and more. *Edmonton Sun*, p. 6.
Hicks, G. (2009c, September 2). Arena riddle a tricky knot to untangle. *Edmonton Sun*, p. 6.
Howard, D. R., & Crompton, J. L. (2004). *Financing sport* (2nd ed.). Morgantown, WV: Fitness Information Technology, Inc.
Hudson, I. (1999). Bright lights, big city: Do professional sports teams increase employment? *Journal of Urban Affairs, 21*, 397–407.
Humphreys, B. R. (2001). The myth of sports-led economic development. *Economic Development Commentary, 25*, 34–37.
Johnson, A. T. (2000). Minor league baseball: Risks and potential benefits for communities large and small. In W. C. Rich (Ed.), *The economics and politics of sports facilities* (pp. 141–151). Westport, CT: Quorum Books.

Johnson, B. K., Groothuis, P. A., & Whitehead, J. C. (2001). The value of public goods generated by a major league sports team: The CVM approach. *Journal of Sports Economics, 2*, 6–21.

Johnson, B. K., Mondello, M., & Whitehead, J. C. (2007). What is the value of public goods generated by a National Football League Team: A CVM approach. *Journal of Sport Management, 21*, 123–136.

Jones, T. (2008, December 17). Time to get it done; Bettman pitches benefits of a new downtown arena for Edmonton and its Oilers. *Edmonton Sun*, p. S3.

Jones, T. (2009, September 1). Downtown going uptown? Katz Group's vision is for much more than a new home for the Oilers. *Edmonton Sun*, p. S3.

Kent, G. (2009, September 1). New Oilers arena 'underway'; Team president confirms project not just speculation. *Edmonton Journal*, p. A1.

Kisilevich, B. R. (2009, September 24). New arena's charm will wear off with first traffic jam. *Edmonton Journal*, p. A19.

Klein, A. (2008, December 20). A fine proposal. *Edmonton Journal*, p. A16.

Kotler, P., Haider, D., & Rein, I. (1993). *Marketing places: Attracting investment, industry and tourism to cities, states and nations.* New York: Free Press.

Lamphier, G. (2009a, August 29). Calgary model could pay city's bills for a new arena–if stars align. *Edmonton Journal*, p. E1.

Lamphier, G. (2009b, September 5). New arena feeds rivalry; Katz Group, Northlands contend for $1B prize. *Edmonton Journal*, p. A1.

Lamphier, G. (2009c, September 12). Public cash means public control: Huntley; If taxpayers help fund new arena, Northlands should run it, chairman says. *Edmonton Journl*, p. A2.

Lamphier, G. (2009d, September 26). Katz Group oddly silent as arena questions pile up; Project appears to lack private-sector partners and developer interest. *Edmonton Journal*, p. El.

Lamphier, G. (2009e, September 29). Katz Group fires back some answers to 'snide' arena questions. *Edmonton Journal*, p. E1.

Lamphier, G. (2009f, October 17). Two cities, two arenas, one big difference; Quebec City's mayor wants governments to pick up 90% of the $400M tab for a new hockey barn. *Edmonton Journal*, p. El.

Lamphier, G., & Kent, G. (2009, August 28). Billion-dollar makeover for core; Oilers owner's arena project could revitalize downtown, sources say. *Edmonton Journal*, p. A1.

Landry, F. (2008, December 14). Mayor mum on plan; Funding scheme for downtown arena expected in new year. *Edmonton Sun*, p. 2.

Landry, F. (2009, October 30). Councillor urges action on downtown roadway makeover. *Edmonton Sun*, p. 11.

Laws, E., Scott, N., & Parfitt, N. (2002). Synergies in destination image management: A case study and conceptualization. *International Journal of Tourism Research, 4*, 39–55.

Letters to the Editor Column. (2009, April 17). *Edmonton Sun*, p. 10.

Loome, J. (2009, September 2). Huge projects warrant plebiscite. *Edmonton Sun*, p. 14.

Mah, B. (2007, February 22). Plan for an aging Rexall. *Edmonton Journal*, p. A3.

McKeen, S. (2008a, December 15). Advice to Oiler owner: Keep arena promises and get a haircut. *Edmonton Journal*, p. A10.

McKeen, S. (2008b, December 22). Fabled columnist looks into his Christmas stocking, and smiles; A fake pearl of wisdom from Gary Bettman, a lump of coal fromWall Street and a chance to volunteer for Operation Red Nose. *Edmonton Journal*, p. A11.

McKeen, S. (2009a, February 27). A cautionary tale for supporters of the downtown airport; Voters expect councillors to put public interest above politics. *Edmonton Journal*, p. B1.

McKeen, S. (2009b, September 2). Katz's arena plan should go ahead, but with strict conditions. *Edmonton Journal*, p. B1.

McKeen, S. (2009c, November 4). Katz's learning curve a steep one; Oilers owner seems to realize his arena dream is complicated and time-consuming when City Hall is your partner. *Edmonton Journal*, B1.

Mackinnon, J. (2009a, October 9). Public perception new arena roadblock; Broad vision required for city's future *Edmonton Journal*, p. C1.
Mackinnon, J. (2009b, October 24). Attitude too much for downtown arena project; Buy-in by population critical to success. *Edmonton Journal*, p. C1.
Miles, M., & Huberman, M. (1994). *Qualitative data analysis: An expanded source book* (2nd ed.). Sage: London.
Noll, R. G., & Zimbalist, A. (1997). The economic impact of sports teams and facilities. In R. G. & Zimbalist, A. (Eds.), *Sports, jobs, and taxes: The economic impact of sports teams and stadiums* (pp. 55–91). Washington, DC: Brookings Institution Press.
Petersen, D. C. (2001). *Developing sports, convention, and performing arts centers* (3rd ed.). Washington, DC: Urban Land Institute.
Platt, M. (2009a, September 1). Cowtown hockey fans drooling with arena envy. *Edmonton Sun*, p. 23.
Platt, M. (2009b, September 1). Yearning for a revamp; CityCalgary fans can only long for arena Edmonton hopes to build. *Calgary Sun*, p. 2.
Pryer, J. (2009, September 2). Who will pay to build it? *Edmonton Journal*, p. A18.
Reporter expresses "regret" for "Deadmonton" article. (2001, August 10). CBC News Online.
Reese, S. D. (2001). Prologue—framing public life: A bridging model for media research. In S. D. Reese, O. H. Gandy, & A. E. Grant (Eds.), *Framing public life: Perspectives on media and our understanding of the social world*. London: City Lawrence Erlbaum Associates.
Riess, S. A. (2000). Historical perspectives on sports and public policy. In W. C. Rich (Ed.), *The economics and politics of sports facilities* (pp. 13–52). Westport, CT: Quorum Books.
Rock, C. P. (2001, April). The case against public subsidies for a new Florida Marlins stadium. *James Madison Institute Policy Report* #31. mimeo
Rosentraub, M. S. (1997a). *Major league losers: The real cost of sports and who's paying for it.* New York: Basic Books.
Rosentraub, M. S. (1997b). The myth and reality of the economic development from sports. *Real Estate Issues, 22*, 24–29.
Saidel, J. (2000). Stadium overview: Myths and realities [On-line]. Available at: http://www.libertynet.org/citycont/overview.htm
Santo, C. A. (2007). Beyond the economic catalyst debate: Can public consumption benefits justify a municipal stadium investment? *Journal of Urban Affairs, 29*, 455–479.
Sassen, S. (2002). *Global networks, linked cities*. London: Routledge.
Shropshire, K. L. (1995). *The sports franchise game: Cities in pursuit of sports franchises, events, stadiums, and arenas*. Philadelphia: University of Pennsylvania Press.
Sidlow, E. I., & Henschen, B. M. (1998). Major League Baseball and public policy, or, take me out to the ballgame, wherever the game may be. *Policy Studies Review, 15*, 65–88.
Simons, P. (2009, September 15). Downtown arena site looks like very tight squeeze; We should consider impact on core area. *Edmonton Journal*, p. A12.
Sound Off!, (2008, December 18). *Edmonton Journal*, p. B3.
Spirou, C., & Bennett, L. (2003). *It's hardly sportin': Stadiums, neighborhoods, and the new Chicago*. Dekalb, IL: Northern Illinois University Press.
Staples, D. (2009a, September 6). Multimillion-dollar faceoff; The debate over a new arena for the Edmonton Oilers has now begun in earnest, raising questions about cost, spinoffs, identity and, most important, who pays. *Edmonton Journal*, p. E4.
Staples, D. (2009b, September 6). Casino may fund arena; Pittsburgh Penguins' new development offers possible model for Oilers: LaForge. *Edmonton Journal*, p. A1.
Staples, D. (2009c, September 20). Time to say goodbye?; It was 35 years ago that Rexall Place opened its doors as Northlands Coliseum. Supporters say there are good reasons to keep it going, starting with tradition. But others say Rexall is too small and dowdy for the modern era of professional sport. *Edmonton Journal*, p. E1.
Staples, D. (2009d, September 21). If levy is used to fund a downtown arena, city will have "big time say," Mandel tells CBC. The Cult Of Hockey Blog.

Staples, D. (2009e, September 23). Big arena must have tax bucks; Northlands chairman says numbers don't add up without public support. *Edmonton Journal*, p. A3.

Staples, D., & Gerein, K. (2009, March 7). Airport battle 'critical': Nichols; Oilers chairman takes leave of absence to fight for city facility. *Edmonton Journal*, p. A1.

Staples, D., & Lamphier, G. (2009, September 12). Oilers' revenue demand would shut out Northlands; Money-losing team needs all cash generated by proposed arena, president says. *Edmonton Journal*, p. A1.

Stewart, J. (2009, September 8). Arena will attract rowdies. *Edmonton Journal*, p. A16.

Suddaby, R., Greenwood, R. (2005). Rhetorical strategies of legitimacy. *Administrative Science Quarterly, 50*, 35–67.

Tankard, J. W., Hendrickson, L., Silberman, J., Bliss, K., & Ghanem, S. (1991). *Media frames: Approaches to conceptualization and measurement*. Paper presented to the annual meeting of the Association for Education in Journalism and Mass Communication, Boston.

Taylor, D. A. (2009, September 2). Who will pay to build it? *Edmonton Journal*, p. A18.

Towering vision. (2009, September 26). *Edmonton Journal*, p. A16.

Turner, R. S., & Marichal, J. F. (2000). Exploring politics on the sports page. In W. C. Rich (Ed.), *The economics and politics of sports facilities* (pp. 189–201). Westport, CT: Quorum Books.

Tychkowski, R. (2009, September 2). Oilers like idea of new digs; Veterans unlikely to play in planned arena district. *Edmonton Sun*, p. S3.

Utt, R. D. (1999). Dreams and schemes: The false promise of a sports-based revitalization strategy. Paper presented at the 1999 Conference of the Southern Regional Science Association, Richmond, Virginia, April 17. mimeo.

Van Diest, D. (2009, September 1). Blueprint for success?; Columbus Blue Jackets' home could be template for new downtown district. *Edmonton Sun*, p. S4.

Van Gorp, B. (2007). The constructionist approach to framing: Bringing culture back in. *Journal of Communication, 57*, 60–78.

Venting. (2008, December 18). *Edmonton Journal*, p. B2.

Venting. (2009a, June 4). *Edmonton Journal*, p. B2.

Venting. (2009b, September 3). *Edmonton Journal*, p. B2.

Venting. (2009c, September 5). *Edmonton Journal*, p. B2.

Venting. (2009d, September 12). *Edmonton Journal*, p. B12.

Virag, T. (2009, March 7). StateCal knows aviation; Nichols leaves Oilers hockey club to lobby for City Centre Airport. *Edmonton Sun*, p. 3.

Waitt, G. (1999). Playing games with Sydney: Marketing Sydney for the 2000 Olympics. *Urban Studies, 36*, 1055–1077.

Warnica, R. (2009, August 29). Arena could leave poor out in cold, advocates say; Development has potential to help or hinder social equality: Inner city groups. *Edmonton Journal*, p. B1.

Whitelegg, D. (2000). Going for gold: Atlanta's bid for fame. *International Journal of Urban and Regional Research, 24*, 801–817.

Zimbalist, A. (2003, December). Sports and economics—A conversation with Andrew Zimbalist. *US Society & Values*, p. 32–35.

Zimmerman, D. (1997). Subsidizing stadiums: Who benefits? Who pays? In R. G. Noll & A. Zimbalist (Eds.), *Sports, jobs, and taxes: The economic impact of sports teams and stadiums* (pp. 119–145). Washington, DC: Brookings Institution Press.

Zipp, J. (1996). The economic impact of the baseball strike of 1994. *Urban Affairs Review, 32*, 157–185.

A Complex Network Approach to Crisis Recovering in Sport Applications

Francesco Carlo Morabito

Abstract Network-based approaches may yield a powerful way to govern complex economic systems like the ones faced in sports, and, particularly, in football. The network approach should be pursued to identify which are the appropriate network structures and topologies that can improve robustness to economic crisis, for example, by facilitating integration and avoiding undesired synchronization events. This requires primarily the modelling of the interactions of structures and dynamics in complex networks that, in turn, strongly encourages a cross-disciplinary fertilization in order to apply models already accepted in other fields and whose impact on real-world problems have been evaluated. In fact, there is a lack of data adequately describing economic networks, nowadays built on intricate interdependencies, trade relations and supply chains on a worldwide scale. In this paper, we try to draw a general picture of what is implied by the complex network approach aiming to suggest models that can both reduce the systemic risk of cascading failures and facilitate the recovery from crisis.

1 Introduction

In terms of dynamical systems, crisis can be interpreted as an undesired consequence of synchronized behavior: a presumably hidden pattern of strong interactions between agents in the financial market is responsible for crashes, in which the interactions are strengthened and the typical parameters measuring network complexity converge to unexpected levels (e.g., the average length is reduced and the underlying tree is packed). However, the existence of a highly networked financial market can be also a potential way to limit the effect of shocks or else to escape from crisis. In order to design a possible intervention to overcome crisis constraints, it is needed to gain an understanding of the non linear dynamics and the evolutionary paths that represent the framework of interactions among different actors. In this paper, we

F.C. Morabito (✉)
University Mediterranea of Reggio Calabria – DIMET, Reggio Di Calabria, Italy
e-mail: morabito@unirc.it

S. Butenko et al. (eds.), *Optimal Strategies in Sports Economics and Management*,
DOI 10.1007/978-3-642-13205-6_7,

shall review the main ideas behind complex networks and graphs, aiming to suggest the application of these concepts to the analysis of economic and financial crisis finally aiming to propose the use of the described methodologies to build instruments that may help facing the crisis: the case of sports activities will be discussed. In our opinion, there is quite a subtle similarity in the study of the dynamic brain pattern formation and emergence of particularly relevant states and the evolution of the economic market before crisis. In particular, we may see the economic crisis as the onset of epileptic seizures (crisis) in the mammalian brain. This event is the result of an anomalous synchronization of neuronal groups initially in a focal area of the brain and then recruiting different, also far, areas of the cortex in a sort of progressive entrainment that end with the crisis. Some authors (see Parlados et al.) suggest that the epileptic seizure is basically a mechanism of resetting the brain after which the normal working operations can be recovered. Similarly, the market crisis can be seen as a long time evolving operational scenario where the crisis is, somewhat, necessary to equilibrate some parameters that cannot be recovered through a linear behavior. There is a need, in this sense, of studying the pattern of intricate connections in the network economy in a dynamic system framework in order to understand the genesis of recruitment processes which give rise to the crisis shock. The current economic crisis dramatically illustrates the necessity of a fundamental understanding of both structure and dynamics of economic networks. Economic systems are obviously built on interdependencies, implemented through trans-national credit and investment networks, trade relations, or supply chains which are very difficult to interpret. It is thus mandatory to highlight an approach that can be able to stress the systemic complexity of this kind of networks and that can be used to revise and/or update established paradigms in economic theory. This could ultimately facilitate the design of policies that may reduce conflicts between individual interests and global efficiency, thus decreasing the risk of global failure by making economic networks more robust. Here, a new framework for modelling time-varying economic networks is analyzed. The nodes in the network may represent firms, societies or companies whereas the directed links represent interactions between them. Both nodes and links evolve according to their own dynamics and on different time-scales. The model assumes that firms' actions take account of the knowledge exchange with other firms, which involves both costs and benefits for the participating actors. In order to increase their utility, companies follow different strategies to create and/or to delete links with other firms. Within this framework, we may find the emergence of different network structures. While many *scale-free* (SF) networks have been introduced recently for describing complex systems, most of them are simple binary random graphs. Here we argue about weighted SF networks, that have been shown to be particularly interesting for the emergence of synchronized features that are, in turn, at the basis of the cascading failures giving origin to crisis. In the next section, we will review some relevant aspect of complex networks, then we describe the network economy. In Sect. 5, we propose to apply these concepts to the sport application of the football societies which interact among them immersed in the general financial crisis. After the conclusions, we will resume the main reference papers from which the ideas here proposed come from.

2 Complex Networks: A Short Review

Complex networks are ubiquitous both in nature and in man-made structures. There exist many excellent reviews and fundamental works on this subject; some of them are reported in the references and concern not only the description of topological features of the underlying graphs but also the characterization of the growth, evolution and dynamics of the networks. The adjective "complex" mainly refers to the consideration of networks composed of a very large number of nodes and links (edges) connecting them. The underlying architecture is governed by shared organizing principles. Most of the interesting processes that can be studied starting from the complex network approach, like robustness, tolerance to node failures and vulnerability to deliberate attacks to a particular class of nodes or edges (for instance, the highly connected nodes) as well as the analysis of synchronization processes have certainly benefited from the recent advances in the knowledge of complex topologies. Some emergent properties of complex networks (being them neuronal networks, the web or the internet, social networks, transportation networks or metabolic ones) are now more understood and novel avenues of researches have been opened. In this short review, we wish to report just the main basic concepts of complex networks. The renowned interest in graphs' studies started from the concept of *random graphs* proposed in a milestone paper in 1960 by the Hungarian mathematicians P. Erdos and A. Renyi. The concept of graph (also directed and weighted) is explained in Fig. 1.

Today, we use the Erdos-Renyi (ER) random graph as a kind of benchmark-to compare with non-random graphs. An historical timeline of significant events in network science can be found in the novel book of T. Lewis. In the late 1960s and 1970s graph theory was used by social scientists to model social networks and study the behavior of humans in groups. Stanley Milgram is credited with introducing the notion of small-world network to the social science community. The experiment of Milgram known as "six degrees of separation" suggested that the distance between

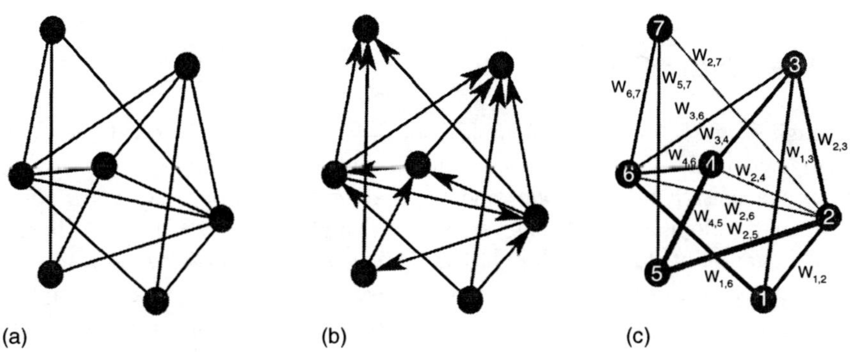

Fig. 1 Undirected (**a**) directed (**b**) and weighted undirected (**c**) graph with $N = 7$ nodes and $K = 14$ links. In directed graphs, adjacent nodes are connected by *arrows*, connecting a source and a sink nodes. In the weighted graph, the w quantities indicate the intensity of the connection, represented by the thickness of the link. From S. Boccaletti et al. (2006)

two people selected at random from the entire population of the United States is approximately six intermediaries. In the late 1990s, a number of scientists in other fields began to use networks as models of physical and biological phenomena. In particular, the pioneering works of D. Watts, S. Strogatz, and A.-L. Barabasi stimulated renewed interest in the theoretical analysis of the networks. Watts compared the structure of very sparse networks with small diameter (*small worlds*) with a diverse number of phenomena such as phase transitions in materials, evolution and metabolism of biological organisms, and distribution of energy in electrical power grids. How could a simple graph model explain such diversity of real-world behaviors? Strogatz studied the impact of network structure on complex adaptive systems in physics as well as explaining why hearts beat in a regular synchronized pattern in mammals, and why certain species of fireflies rhythmically synchronize their chirps in unison without centralized control. It appeared that living organisms tend to synchronize their behavior without global rules, but in an *emergent* way. Synchronization is a by-product of the structure of "living networks." Barabasi introduced the concept of *scale-free* free networks, that can be defined a kind of non-random networks with hubs. In a number of studies about the structure of the Internet and of the World Wide Web (WWW), Barabasi et al. discovered an emergent property of the decentralized Internet, that means that it had emerged without central planning into a structure consisting of a small number of extremely popular sites called hubs. Instead of being random, like an ER (Erdos-Renyi) network, the Internet topology was highly non-random. In fact, the probability that a site has k links obeys a power law, which drops off quickly for large k. Furthermore, they speculated that this was the result of a rule of growth and evolution of the underlying graph which was named preferential attachment, meaning that the probability a site will obtain a new link is directly proportional to the number of links it already has. Thus, the more links a site has, the more it gets, the so-called "rich get richer" phenomenon. In the following sub-sections, we shall give some relevant additional element needed to understand the concept of scale-free networks and the dynamical phenomena of robustness, vulnerability and synchronization.

2.1 Some Basic Concepts

Although many definitions of complex networks have been proposed and deeply investigated in the last decades, three major concepts, namely, the *average path length*, the *clustering coefficient*, and the *degree distribution*, play a key role in the recent developments and achievements of complex networks theory. In fact, the original attempt of Watts and Strogatz in their work on *small-world* networks, was to construct a network model with small average path length as a random graph and relatively *large* clustering coefficient as a regular lattice, which evolved to become a new network model. On the other hand, the discovery of *scale-free* networks was based on the observation that the degree distributions of many real networks have a power-law form, although power-law distributions have been largely known for a long time in physics, for many other systems and processes.

2.1.1 Average Path Length

In a network, the distance d_{ij} between two nodes, labelled i and j respectively, is defined as the number of edges along the shortest path connecting them. The *diameter D* of a network, therefore, is defined to be the maximal distance among all distances between any pair of nodes in the network. Accordingly, the *average path length L* of the network is defined as the mean distance between two nodes, averaged over all pairs of nodes. Here, L determines the effective "size" of a network, the most typical separation of one pair of nodes therein. In a friendship network, for instance, L is the average number of friends existing in the shortest chain connecting two persons in the network. It was an interesting discovery that the average path length of most real complex networks is relatively small, even in those cases where these kinds of networks have many fewer edges than a typical globally coupled network with an equal number of nodes. This is reported as the *small-world effect*, hence the name of *small-world networks*.

2.1.2 Clustering Coefficient

If we observe a *social network*, it is expected that your friend's friend is also your direct friend; namely, two of your friends are quite possibly friends of each other. This property refers to the *clustering* of the network. It is possible to define a *clustering coefficient C* (also known as *Cluster Index*) as the average fraction of pairs of neighbors of a node that are also neighbors of each other. Suppose that a node i in the network has k_i edges and they connect this node to k_i other nodes. These nodes are all neighbors of node i. Clearly, at most $k_i(k_i - 1)/2$ edges can exist among them, and this occurs when every neighbor of node i is connected to every other neighbor of node i. The clustering coefficient Ci of node i is then defined as the ratio between the number E_i of edges that actually exist among these k_i nodes and the total possible number $k_i(k_i - 1)/2$, namely, $C_i = 2E_i/(k_i(k_i - 1))$. The clustering coefficient C of the whole network is the average of C_i over all i. Clearly, $C < 1$, and $C = 1$ implies that the network is globally coupled, which means that every node in the network connects to every other node. In a completely random network consisting of N nodes, C is proportional to $1/N$, which is very small as compared to most real networks. It has been found that most large-scale real networks have a tendency toward clustering, in the sense that their clustering coefficients are much greater than O(1/N), although they are still significantly less than one (namely, far away from being globally connected). This, in turn, means that most real complex networks are not completely random.

2.1.3 Degree Distribution

A single node is most simply characterized (and, somehow, ranked) by its *degree*. The degree k_i of a node i is usually defined to be the total number of its connections. Accordingly, the larger the degree, the "more important" the node is in a network. The average of k_i over all i is called the *average degree* of the network, and is denoted by $< k >$. The spread of node degrees over a network is characterized by a distribution function P(k), which is the probability that a randomly selected node has exactly k edges. A regular lattice has a simple degree sequence because all the nodes have the same number of edges; and so a plot of the degree distribution contains a single sharp spike (delta distribution). Any randomness in the network will broaden the shape of this peak. In the limiting case of a completely random network, the degree sequence obeys the familiar Poisson distribution; and the shape of the Poisson distribution falls off exponentially, away from the peak value $< k >$. This exponential decline implies that the probability of finding a node with k edges becomes negligibly small for $k >>< k >$. In recent years, many empirical researches showed that for most large-scale real networks the degree distribution deviates significantly from the Poisson distribution. In particular, for a number of networks, the degree distribution can be better described by a *power-law* of the form $P(k)\ k^{-\gamma}$ (typically, the exponent $2 < \gamma < 3$). This power-law distribution falls off more gradually than an exponential one, allowing for a few nodes of very large degree to exist.

Figure 2 shows the difference existing between a power-law and an exponential distribution: the power-law distribution has a heavy tail. Since these power-laws are free of any characteristic scale, such a network with power-law degree distribution is called a scale-free network. In non-mathematical terms, a *scale-free network* is one with a small number of nodes showing a high-degree and a a large number of nodes with low-degree. The rare nodes with high degree are called *hubs*; therefore,

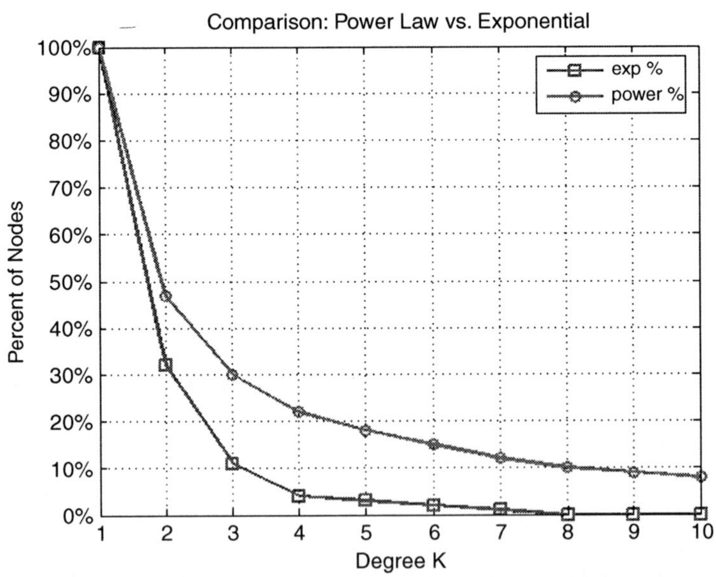

Fig. 2 Nodes with Degree k, in percent, versus the Degree k for Random and Scale-Free Graphs (power-law distribution). The SF networks includes some hubs nodes with a very high number of links, although most nodes have just a few connections. If plotted in a double-logarithmic scale, the distribution of nodes results in a straight line

scale-free networks are networks with hubs, which results in a skewed degree sequence distribution. It is a fact that small-world and scale-free features are common to many real-world complex networks. Many networks have a *community* structure, in that case, nodes are linked together in densely connected groups between which connections are sparser.

2.1.4 Weighted Networks

Real networks, like the ones analyzed in this contribution, are not binary, where the edges between nodes are either present or not; on the contrary, they display a large heterogeneity in the capacity and the intensity of the connections. Simple examples are the existence of ties between individuals in social networks, or the synapses electrical signals in neural networks. In these networks, each link is associated to a numerical value measuring the strength of the connection. Weighted networks are characterized in terms of statistical quantities. Standard graphs are characterized by some matrices that implement the map of connections among the nodes. The more used is the *adjacency matrix*, that assumes for each entry $a_{ij}=1$ if i and j are connected, otherwise $a_{ij} = 0$. In weighted networks, the standard adjacency matrix is substituted by a weight matrix showing not only if the link is present but also the intensity of the link. Asymmetrically weighted networks, where $w_{ij} \neq w_{ij}$ have special properties in enhancing synchronization. A simple characterization of weighted networks can be obtained by the weight distribution Q(w), namely the probability that a given edge has weight w. The node strength generalizes the concept of node degree and integrates the information on the number (degree) and the weights of links incident in a node. In a weighted network, the path joining two nodes with the minimum number of edges (shortest path) is not necessarily the optimal path. It is, thus, useful to define a different weighted shortest path length.

2.1.5 Special Properties of Complex Networks

Apart from the concept of evolution of networks that strongly differentiate the analyzed models from standard static networks, the discussed topologies show some relevant properties. The SF networks are claimed robust to avoid a malfunctioning when a fraction of its components is damaged. This property largely affects the efficiency of any process running on the network.

Figure 3 describes the different abilities of ER random graphs and BA scale-free models with respect to this property, in terms of both the average shortest path length, L, and the global efficiency of the damaged network, E. A large number of nodes (up to 80%) are removed. Since the network tends to become unconnected, E is a better quantity than L to describe the system. Here, it is clearly seen the double face behavior of scale-free networks: they are robust to errors but they strongly suffer any deliberate attacks. In Fig. 4, it is reported the level of efficiency of a very large network after the system has reached a stationary state following dynamic damage. In the simulation, taken from Crucitti et al. instead of permanently

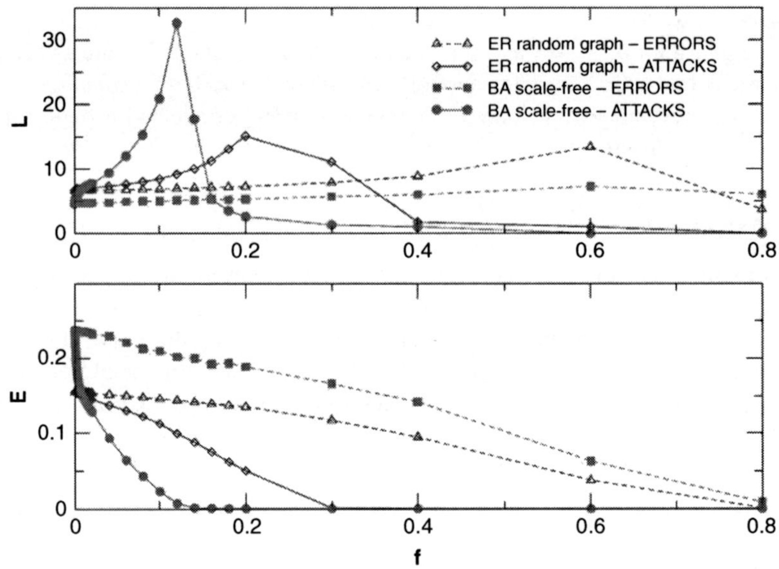

Fig. 3 Robustness of the network under random failures and attacks on nodes. The average shortest path length L and global efficiency E are reported as a function of the fraction of removed nodes f. An ER random graph and a scale-free network, both with $N = 5{,}000$ (*nodes*) and $K = 10{,}000$ (*links*) are considered. From Crucitti et al. (2003)

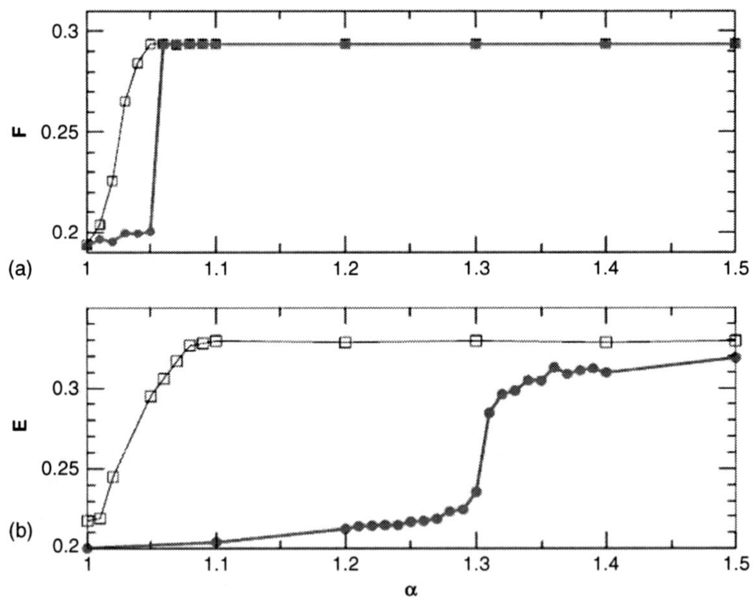

Fig. 4 Cascading failures in (**a**) ER random graph and (**b**) scale-free network as triggered by the removal of a node chosen at random (*squares*), or by removal of the node with largest load (*circles*). Both the networks considered have $N = 2{,}000$ and $K = 10{,}000$. The efficiency, E, of the network is reported as a function of the tolerance parameter? In the case triggered by the removal of a node chosen at random the curve corresponds to an average of 10 triggers. From Crucitti et al. (2004)

removing the overloaded nodes, the communication through them is gracefully degraded, thus the flow of information (energy) will avoid that kind of nodes. The model assumed that the underlying graph is weighted. Each weight, $0 < w_{ij} < 1$, measures the efficiency in the communication along the i–j edge. At the initial time (say, $t = 0$), $w_{ij} = 1$ for all the links, meaning perfect communication. The removal of some nodes starts the redistribution dynamic process along the network: there is a change in the most efficient paths between nodes and thus the distribution of the loads, creating overload on some nodes. Because of the efficiency's reduction of some links, the flow will take some alternative paths. The damage caused by cascading failures is measured by the decrease in the network efficiency. For both ER and BA scale-free models, we observe a decrease of the efficiency for small values of the tolerance parameter α, and the collapse of the system for values smaller than a critical value α_c. Two different triggering strategies (random and load-based damaging) have been considered. In both kind of networks the collapse transition is sharper for load-based removals. As previously discussed, ER networks have an exponential load distribution while BA networks exhibit a power-law distribution in the node load. In the latter models, there are few nodes, the ones with extremely high initial load, that are far more likely than other nodes to trigger cascades. This is clearly seen in Fig. 4b, that shows the existence of a large region in the tolerance parameter where scale-free networks are stable with respect to the random removal and unstable with respect to the load-based removal. This observation implies that, in most cases, the network is generally stable to an initial shock related to the breakdown of a node: the failure is tolerated and reabsorbed by the whole system. There is, however, a finite, although small, probability that a failure triggers an avalanche mechanism collapsing the whole network. This kind of behavior of the scale-free networks is truly relevant for our studies, since they are the most typical real models we meet in the network economy.

2.1.6 Synchronization in Complex Network

One of the most recently investigated aspects of complex networks is certainly the emergence of collective and synchronized dynamics. Recently, synchronization in different small-world and scale-free dynamical network models has been carefully studied. These studies may shed new light on the synchronization phenomenon in various real-world complex networks. Some researches are focusing on assessing the propensity for synchronization of a network: the main result that was achieved is that optimal topologies in the coupling configuration may provide enhancement of the synchronization abilities. As an example, in complex networks, it was shown that random long-range connections lead to a more rapid and robust phase locking of nonlinear oscillators than nearest-neighbor coupling or locally dense connection schemes. The onset of synchronization is ruled by an order parameter, typically the coupling strength between pairs of connected oscillators, r. For values of the order parameter greater than a critical one, the network tends to synchronize. Given the dynamics of an isolated node and the linking structural matrix, the synchronizability of the dynamical network with respect to a specific coupling configuration is said to be strong if the network can synchronize with a small value of the coupling strength r. Recent results have shed light on the influence of the topology of the local interactions on the route to and the onset of synchronization. The results suggest that the onset of synchronization is mainly determined by the clustering coefficient, C, namely, networks with high C promote synchronization at lower values of the coupling strength. On the other hand, when the coupling is increased beyond the critical point, the effect of the average shortest path length dominates over C and there is a delay effect in the appearance of the fully synchronized state as the average shortest path length increases. Other recent results showed that the presence of weighting connections may have a strong impact on synchronization. The main ingredients of this improved propensity to synchronize are as follows: (1) the weighting must induce a dominant interaction from hub to non-hub nodes; for example, in a star network consisting of a single large hub, when the center drives the periphery of the star, synchronization can be easily achieved; (2) the network contains a structure of connected hubs influencing all the other nodes; there is a need of dominant interactions from hubs to non-hubs for synchronization be more likely. In the next section we will see that in *scale-free* networks the weighting procedure based on link loads enhance the propensity for synchronization of the networks.

3 Network Economy

Recently, the huge growth of interest and the renaissance of the studies on networks and graphs motivated new findings about the power of networks in Economy from the organization of companies to marketing. The main example proposed by Barabasi in his fundamental book, regards the formation of a sparse network of a few powerful managers controlling all major appointments in Fortune 1,000 companies. The growth of a network of alliances is the basis of success in the bio-tech industry, the structure and the underlying topology of the network is the key to gain an organization's ability to adapt to rapidly changing market conditions. Also, the ideation of strategies based on the essence of complex networks can lead to unpredictable successes in marketing. In other word, understanding the network effects, in terms of dynamical growth and preferential attachments to hubs is needed to guarantee a survival in the new economy evolution. The standard organization of firms was long time based on a hierarchical structure, i.e. a tree with decreasing responsibility levels starting from the root till the final executors of orders derived from it. The tasks of manager were non-overlapping and the work became specialized through the bifurcating branches. The structure of tree shows problems in terms of information flow that should be properly filtered in order to avoid overload in inappropriate branches; however, the main limitation stems from the rigidity of the organization which can be very stiff: each possible problem and/or implemented modification in the productive cycle could strongly impact on all levels of the tree generating shut down and inability to recovery and to adapt to changing environment. Finally, the firm's model relying on hierarchy is best suited for mass production, where, in a sense, the hardware continuously improved is the main productive factor. Today, in the information era, the assets are more volatile and based on ideas, exchange of information and preparation for change. Flexibility of organization calls for a rethinking of the entire structure for facing the requests of the information economy. For example, this gave rise to outsourcing, i.e. the practice of renouncing possibly to the industrial secrets and giving out some manufacturing parts of the production cycle. The major step of the organizational change is the shift from a tree to a network structure: the hallmark of the complex graph topology are the web-map of directed links between the nodes. Sometimes, the integration is just virtual and the reach of business expands from local (domestic) to global (worldwide). The invasion of new markets with novel products requires both a political level of interactions among integrated countries (like European Union) and new alliances between firms, apart from internal parts of the same company. In short, companies change their organization to survive in a fast changing economic world from the optimized tree where each actor has a specific, truly specialized, role to a dynamic and continuously evolving, more flexible structure. A strong consequence of this novel approach to market is the search for collaborations with other institutions, for example by inclusions of the top managers in other organizations' board of directors. The interactions are the effect of the network structure. A short group of top managers is involved in several boards taking decisions in important steps for companies' future, like merging, fusions, and acquisitions. In the network generated by the directors, each node is a member of the boards linked to other directors: being thousands of firms so linked through their directors, the result is a complex web of relationships with the character of a small-world network due to a limited but not to be neglected number of directors who serve on several boards. According to Barabasi, with reference to the network of Fortune 1,000 companies, made up of 10,100 directorships held by 7,682 directors, well 14% of directors serve on two boards and about 7% serve on three or more. The resulting 21% of directors implement a small-world network with five degrees of separation. These intricate relationships among board directors form the underlying topology of the economic network; however, to really gain insights into the dynamics developing, it is needed to explain the interactions among agents. The alliances among corporations make the network emerge: some companies are related through a high level of connectedness that is not possible to achieve in both random and small-world networks. In these kind of networks, some nodes work as hubs, which are responsible for leaving the complex network connected: these are companies with a large number of partnerships. Through the hubs, like in a network of airports, it is possible for not directly connected nodes to maintain relationships. In particular, for the bio-tech industries, dynamically related by exchanging knowledge and patents in $R\, and D$ partnerships, it was shown that the evolution of the network is ruled by a scale-free scheme. This means that the number of companies that entered in collaboration with exactly k other firms, representing the number of links they form in the network, followed a power-law, the main property of a scale-free network. In the case of sports like football, the scheme is not so different: we have a group of well-connected large societies that work as hubs for a large number of small societies, all of them integrated into an evolving scale-free network. This topological structure works well in normal times, but is strategic during and by recovering from crisis. One of the characteristic of football societies is that the alliances are fluid, periodically renegotiated as the consumer interests and the evolving marketplace shift: we can use this case as a paradigm of future worldwide network economy. It is quite strange that these obvious network effects are neglected from economic theory. In the standard formal model, the economy is seen as the result of the interaction of anonymous individuals through the price system: in this framework, the actions taken by the individual agent (companies and consumers) have little consequence on the global market. Actually, the market is by itself a network where companies, firms and institutions are the nodes and the interaction among the agents are the links (or branches) of the underlying graph. What's more interesting in the network economy is that the (directed) links are weighted through a sort of "value" of the transaction. The preferred model for studying the evolution of the network is not a binary one, but a real network displaying a large heterogeneity in the capacity and the intensity of the connections. As previously discussed, a weighted network is the theoretical scheme to be used. It comes out that the a weighted network has a special propensity for synchronization, i.e. the presence of weighted connections has relevant consequences in determining the network's dynamics. Several authors studied how the presence of weighted connections can enhance the propensity to synchronize. It has been shown that this can be reached by making use of the information contained in the global topology. The main idea is to scale the coupling strength between two nodes to the load of the edge connecting them. The load l_{ij} of the link connecting nodes i and j is a measure of the traffic of shortest paths including that link. This can be interpreted as a measure of the network structure at a global scale. By counting the number $n(i, j)$ of the shortest paths connecting the nodes i and j for all the couple of nodes in the network, one can scale the load by $1/n$, this way associating to the load distribution the information on the network structure of pathways at a global level: indeed, the value of each load l_{ij} can be influenced also by pairs of nodes that are far away from the nodes i and j. This procedure of weighting does not use the information on the node degrees since nodes with high degrees can have links with low load and, on the contrary, nodes with low degree may be linked by connections with high load. It was shown that for networks weighted based on link load the propensity for synchronization is by far improved with respect to weighting procedure based on the node degree distribution. This is particularly true for scale-free topologies. In the network economy, by consequence, the synchrony effect is favored and could give rise to catastrophic failures, as reported by Barabasi with reference to the case of the cascading failures of companies and financial institutions in Thailand, Indonesia, Malaysia, Korea ant the Philippines starting from the failure of Somprasong Land, a Thai property development company, on February 5, 1997. The Asian crisis was a large-scale example of a cascading financial failure: these kind of events cannot be interpreted within the framework of the standard market model, but they are direct consequence of the interdependencies generated in an economy network-structured. Understanding these events can help to limit future crisis, by some anticipated intervention on the nodes and the paths of the damage.

4 Crisis in the Framework of Complex Networks

The economic crisis we are facing today is paradigmatic of a critical need for a deep understanding of the topological structure and dynamics of economic networks and of the importance of novel investigation on these topics. Economic systems are increasingly built on interdependencies, implemented through trans-national credit and investment networks, trade relations, or supply chains that have proven difficult to predict and control. As clarified in the previous section, it is needed an approach that may highlight the systemic level and complexity of economic networks that should drive novel researches and paradigmatic developments in the economic theory. This is a privileged way to facilitate the future strategies and policies in order to reduce conflicts between individual interests and global efficiency: the goal to be reached is a reduced systemic risk of global failure and, thus, the gain in robustness of economic networks. Indeed, the current crisis is hardly predictable by considering the failure of a few relevant agents. The current view is to relate the system's dynamics to the topological properties of the underlying complex network. The two approaches to the studies of economic networks can be resumed as follows: (i) a molecular level, in which the focus is on individual agents and the map of their relationships; (ii) a macro-level, where the perspective looks to some statistical regularities of the network considered as a whole. Both the approaches have some advantages and disadvantages: at a molecular level, we emphasize the incentives of agents in developing links within nodes, this way failing to understand the global dynamics; at the macro-level, we can learn the large-scale properties of the network at the cost of losing the effective impact of individual agents. To prevent the onset of crisis it is now clear that it is mandatory to gain a deep understanding of not only the dynamics that originate the real economic complex networks but also the evolution at the two-scale level (individual agents—global network) assuming nodes heterogeneity and weighted links (in terms not only of hubs structure and co-operation but also of loads and suitable probabilities of interactions). In fact, as reported in the literature, heterogeneities of agents can give rise to dynamics of the networks that prevent phase transitions, this way introducing a source of unexpected stability. In times of crisis an order parameter of network dynamics is the individual search for efficiency, conflicting with aggregate welfare. The environmental volatility due for example to individual governs ephemeral actions, like temporary incentives to some kind of firms, or to innovation in the productive cycles can strongly reduce the network efficiency measured on the basis of the aggregate centrality of agents. The environmental volatility implies that when a single agent is exposed to an exogenous shock, it may force the deletion of links. If the environmental volatility grows up at a critical level, the network structure is subject to changes and breaks down into a sparse one, thus reducing the global efficiency. If a node fail, it may induce cascading failure, as previously argued, thus generating the systemic risk. This is certainly the case of football societies, where links represent debts and credits spread over multiple years between linked clubs. In general, a strongly connected networks can more easily compensate for individual failures by forcing diversification; however, this is not true for the failure of a few hubs or the deletion of links with heavy load. Since the removal of group of nodes may result in obtaining less stable networks, it seems that the scale-free structure is best suited to prevent cascading crashes.

5 Sport: A Network of Societies and Consumers

Sport as a social phenomenon per se and as an interesting research topic have gained increased popularity during nineties. Sports, as a leisure activity and, in particular, at a professional level, have also grown into an industry sector of their own: key examples of the increased popularity is the emergent development of sports hall of fame and museum complexes all over the world and the parallel development of marketing centers selling gadgets with the societies marks. The sport industry, including sport events management, can be nowadays regarded as an important and relevant part of the more general entertainment business. Sport, and particularly football societies, tend currently to make partnership to maximize the global utility in terms not only of direct money income, like TV revenues and ticket costs, but also of increment of players' value, society marketing initiatives and other somehow related financial and commercial activities of the top management. In the recent literature, scientific results can be found in terms of both modelling of networks of collaborative agents that maximize the total profit (efficiency of the network) and the cost of adding and deleting topological nodes. It has been shown that this process can drive the network to multiple stable states (dynamic evolution analysis) and there is a relevant interest in trading-off network stability and efficiency. In football, like in the economic networks we described in previous sections, the societies network show the characteristic fat-tail behavior of scale-free networks, which suggests that only a limited number of top societies (in Italy, for example Inter, Milan, Juventus; in Spain, Barcelona and Real Madrid) interact with many others, also outside the country. The big societies form clusters and they are the hubs of the complex network. Here, the links are weighted because of the level of economic interactions under consideration (think, as an example, to the case of the financial trade of changing the top players Ibrahimovic and Eto'o, between Barcelona and Inter, or the transfers of Kakà and Cristiano Ronaldo from Milan and Manchester United to Real Madrid, both in the summer 2009). The weights of the links implies that it is not only important to know if a link between two societies does exist, but it represent, for example, traded volumes, invested capitals, and so on. The evolution of the resulting network is different, as we have discussed, not only in terms of growth and dynamical paths, but what's more, in terms of stability and propensity to synchronization. Definitely, the model we are building strongly affects the onset of cascading failures and, at the same time, in principle, the success in avoiding the impact of crisis and the ability to recovery from it. In the football framework, there are some well-known financial problems that we should look at following the complex network approach we are suggesting in this paper. TV revenues are one of the main financial income for societies. Here the network effect can be easily understood: it is clear that top clubs gain most of the money income from TV, but without small clubs there are no TV game diffusion. This is simply stated in the fundamental Neale's paper, through the "*Louis Schmelling Paradox*" referring to the boxer Joe Louis whose earning were higher if there was an evenly matched contender available for him to fight than if the nearest contender was relatively weak. In professional sport the monopoly is less profitable than competition: the football team is a firm that aims to improve the profits, however, it has to share the market with other teams that co-operate with each other to produce a viable League competition. Contracts of top players for advertising could in part help to mitigate this effect: part of the money for funding players wages can derive from non-football companies that wish to ad their products through the image of top players. Players' wages and transfer fees increases is another aspect to be analyzed. Till some years ago, there was a wage inflation and a sudden increase trend in the top players' wages. This implies limited investment in small societies of lower divisions. This generates not only controversies in the relationships within League's organizing body, but a weakness of the nodes less linked and a reduction of investment in the young players (minor) teams. In European countries, football teams used to spend in advance the money supposed to come from future revenues. On the other hand, the balance sheets are mostly based on intangible fixed assets, that are, by their very nature, mostly volatile. For example, it is not possible to guarantee that the cost of a top player does not decline in future years, since it depends on the market conditions. Also, the TV revenues could change, because of financial world crisis, the team could be relegated to Second Division, and so on. Furthermore, the total debt of most clubs is increasing. In summary, there is a condition of unstable financial equilibrium that could be well understood in order to avoid cascading failures. We have seen that the environmental volatility strongly impact on network's efficiency and that after a threshold value the network structure can face a breakdown effect. If some clubs should not survive to crisis, being exposed to exogenous shocks, in the network model we shall have the deletion of nodes and links. The ephemerality of the value of players, the proportional reduction of tangible fixed assets and the out-of-equilibrium state of some agents (clubs) due to balance sheets weaknesses going towards bankruptcy condition reduce the performance of the network as a whole. Big societies search for a community of alliances with other big clubs to prevent collapses: unfortunately the formation of communities (groups of nodes such that there is a higher density of edges within groups than between them) cannot help stability. The above discussed scale-free structure appears a more appropriate schemes to improve robustness (i.e., the ability of the network to avoid failures when a fraction of its constituents is damaged) in a systems where deliberate attacks to nodes are not predictable. Finally, spending the future income and the structural weakness of club's balance sheets can strongly impact the equilibrium of the networks and generate a route to phase transitions, instability and cascading failures.

6 To Escape from Crisis

In conclusion, it seems clear that a strong effort should be made in better understanding both the dynamic of formation of networks, through nodes attachment and deletion, and the role of network topology in determining paths to synchrony and, thus, economic crisis. It is important to clarify that the knowledge of complex network in economy can yield suggestions in order (a) to favour the emergence of more stable and robust networks (b), to anticipate the onset of cascading failures also motivated by exogenous triggers, like financial crisis; (c) to recovery from crisis by a dynamic reconfiguration of the underlying graph implying the knowledge of the way topology and dynamic evolution interact. The football societies network could present a scale-free structure governed by power-law scaling, due to the evidence of hubs (great clubs, like Barcelona, Real Madrid, Inter, Milan, Chelsea, Manchester United) in different countries. In fact. there is an exchange of players between top clubs and smaller ones in different countries and the number of interactions is growing continuously. The network we analyze is certainly weighted and includes lot of cycles and motifs, thus the propensity to synchronization effects should be deeply investigated (the comparison with cortical network generating hyper-synchronous behavior driving to epileptic seizures is structurally confirmed). It is clear that an effort should be made to obtain more and better quality data in order to simulate the behavior of large-scale economic networks: the definition and generation of properly designed databases reflecting clubs interactions and network properties is to be fostered.

7 Conclusions

In this earlier version of the paper, we have proposed a critical reading of the financial-economic crisis on the structured economic networks. The analysis of sport societies, in particular football ones, can be considered as a paradigmatic case-study for the analysis of the impact of both topology and dynamic interactions among nodes in a network of agents representing the clubs interacting through competitions and Leagues organizations. The model of weighted complex networks has been considered as the most relevant one in order to both understanding the genesis of crisis and cascading failures of nodes and finding suitable ways to reduce the impact of the crisis or to prevent it. A comparison is proposed with the cortical (neuronal) networks in the brain, with particular reference to the hyper-synchronization of sub-networks that generate the epileptic seizures. The paroxysmal activity may be localized or spread by going through cascading failures. The resetting of the previous state and, thus, the restoring of normal brain activity is seen as a dynamical process that emerge on the underlying topology of involved network of networks. Similarly, the collapse of an economic network could be predicted and mitigated by an accurate description of the weighted complex network. In that case, it is possible to early recovery from crisis by a dynamic redistribution of the network load and flow. As a relevant example, it is discussed the case of football teams interacting in a very special example of economic network, subject to decisions which are taken and implemented collectively at league level, by eliminating geographic boundaries and limitations.

References

Albert, R., & Barabasi, A. L. (2000). Topology of evolving networks: Local events and universality. *Physical Review Letters, 85*, 5234–5237.

Arena, A., Diaz-Guilera A., Kurths J., Moreno Y., & Zhou C. (2008). Synchronization in complex networks. *Physics Report*, preprint submitted.

Barabasi, A. L. (2002). *Linked*. Cambridge, MA: Perseus Publishing.

Barabasi, A. L., & Albert, R. (1999, October). Emergence of scaling in random networks. *Science, 286*, 509–512.

Barabasi, A. L., & Bonabeau, E. (May 2003). Scale-free networks. *Scientific American, 288*, 50–59.

Boccaletti, S., Latora, V., Moreno, Y., Chavez, M., & Hwang, D.-U. (2006). Complex networks: Structure and dynamics. *Physics Report, 424*, 175–308.

Calderon, C., Chong, A., & Stein, E. (2007). Trade intensity and business cycle synchronization: Are developing countries any different? *International Journal Economics, 71*, 2–21.

Crucitti, P., Latora V., Marchiori M., & Rapisarda A. (2003). *Physica A, 320*, 622.

Crucitti, P., Latora V., & Marchiori M. (2004). Modeling for cascading failures in complex networks. *Physical Review, E69*, 04510 (R).

Dobson, S., & Goddard, J. (2001). *The economic of football*. New York: Cambridge University Press.

Erdos, P., & Renyi, A. (1960). On the evolution of random graphs. *Publ. Math. Inst. Hung. Acad. Sci., 5*, 17–61.

Gama Oliveira, J., & Barabasi, A. L. (2005). Darwin and Einstein correspondence patterns. *Nature, 423*, 1251.

Garcia, J., & Rodriguez, P. (2002). The determinants of football match attendance revisited: empirical evidence from the Spanish Football League. *Journal of Sports Economics, 3*, 18–38.

Gonzales, M. C., & Barabasi, A. L. (2007, April). Complex networks: From data to models. *Nature, 3*, 224–225.

Iasemidis, L., Sabesan, S., Chakravarthy, N., Prasad, A., & Tsakalis, K. (2009). Brain dynamics and modeling in epilepsy: Prediction and control studies. In S. K. Dana, P. K. Roy, & J. Kurths (Eds.), *Complex dynamics in physiological systems: From heart to brain*. Springer Science + Business Media B.V.

Jeong, H., Tombor, B., & Barabasi, A. L. (2000, October). The large scale organization of metabolic networks. *Nature, 407*, 651–653.

Konig, M. D., Battiston, S., Napoletano, M., & Schweitzer, F. (2008). *The efficiency and evolution of R and D networks* (Economic Working Paper Series). Zurich: ETH.

Lazer, D., Pentland, A., Adamic, L., Aral, S., Barabasi, A. L., Brewer, D., et al. (2009, February). Computational social science. *Science, 323*, 721–723.

Lewis, T. G. (2009). *Network science*. New Brunswick, NJ: Wiley.

Milo, R., Shen-Orr, S., Itzkovitz, S., Kashtan, N., Chklovskii, D., & Alon, U. (2002, October). Network motifs: simple building blocks of complex networks. *Science, 298*, 824–827.

Schweitzer, F., Fagiolo, G., Sornette, D., Vega-Redondo, F., Vespignani, A., & White, D. R. (2009, July). Economic networks: The new challenge. *Science, 325*, 422–425.

Sabesan, S., Chakravarthy, N., Tsakalis, K., Parlados, P., & Iasemidis, L. (2009). Measuring resetting of brain dynamics at epileptic seizures: application of global optimization and spatial synchronization techniques. *Journal of Combinatorial Optimization, 17*, 1, 74–97, Springer Verlag.

Serrano, M. A., & Boguna, M. (2003). Topology of the world trade web. *Physical Review, 68*, 015101.

Sorrentino, F., & Ott, E. (2008). Adaptive synchronization of dynamics on evolving complex networks. *Physical Review Letters, 100*, 114101.

Strogatz, S. H. (2001). Exploring complex networks. *Nature, 410*, 268–276.

Wang, X. F., & Chen, G. (First Quarter 2003). Complex networks: Small-world, scale-free and beyond. *IEEE Circuits and Systems Magazine, 3*, 6–20.

Watts, D. J. (1999). *Small worlds*. Princeton, NJ: Princeton University Press.

Watts, D. J., & Strogatz, S. H. (1998). Collective dynamics of small-world networks. *Nature, 393*, 440–442.

A Study of Fairness in Fourball Golf Competition

Patrick Siegbahn and Donald Hearn

Abstract The golf handicap system is designed to be fair in single player games. The same system is also applied to team games with two or more players per team. In this study the fairness of the particular team golf game known as fourball match play was analyzed as well as measures to improve it. By using historical player data, software was produced to simulate games. The level of fairness was estimated given the handicaps of the players. It was determined that:

- Players with high handicap have high variance and therefore also an advantage.
- Teams should seek to optimize the spread in handicaps by pairing with someone as close to 11 or 29 handicap strokes apart and as far from 0 or 18 as possible.
- The handicap allowance introduced by the United States Golf Association to penalize high handicappers in team golf is not an effective tool to handle the unfairness.
- The tiebreaker rule is a simple way to lower the general unfairness by roughly 70%.

1 Introduction

Golf is one of the few individual sports where competitors do not directly affect their opponent's performance. This feature of the game encourages the use of a handicap system to allow for fair competition between two players of any level. If any golfer's score on 18 holes is considered to be a normally distributed random variable, then the aim of handicapping is to shift the two distributions to center them on the same mean value, see Fig. 1. The golf handicap system has been modified and improved over many decades, with a primary emphasis on golf games in which two players compete against each other. However, handicaps are also used when golf is played as a team game with two or more players per team, and it is common to use the minimum of the individual scores as the team score. Higher-handicapped players

P. Siegbahn (✉)
Royal Institute of Technology, Stockholm, Sweden
e-mail: patrick@siegbahn.com

S. Butenko et al. (eds.), *Optimal Strategies in Sports Economics and Management*, 143
DOI 10.1007/978-3-642-13205-6_8, © Springer-Verlag Berlin Heidelberg 2010

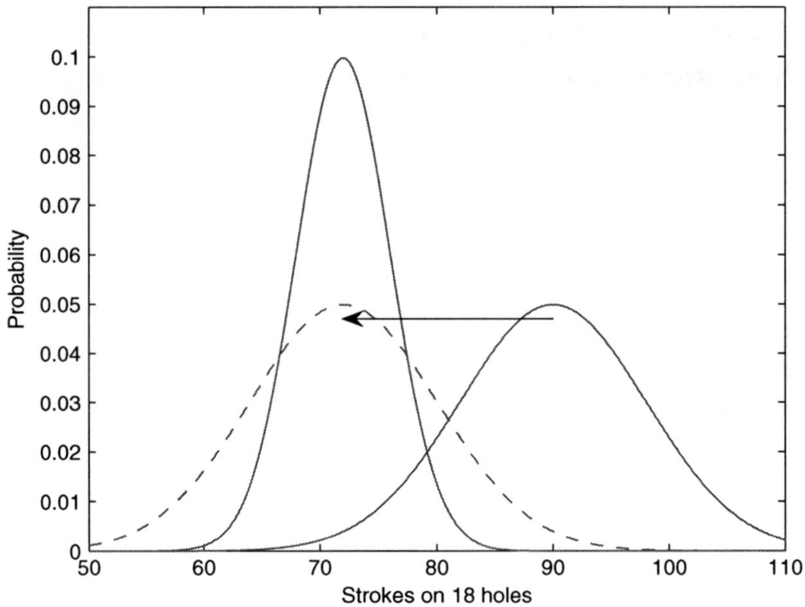

Fig. 1 The purpose of handicapping is to center all players' distributions on the same mean

generally produce a wider range of hole-by-hole scores than better, more consistent players. This means that the team with high handicaps could have an advantage. The purpose of this study is to investigate the issue of fairness in the particular team game known as *fourball match play*.

1.1 History of Handicapping

In 1911 the United States Golf Association (USGA) adopted the first handicap system using the average of the three best rounds of golf (Pope of Slope, 2008). The system was modified over the years to end up with the current method in 1967. In 1976 two additional adjustments were made. Allowances were added to team games and a previous moderation in general handicaps was dampened as compensation. Since the new allowances were not accepted by golfers in general, the USGA decided to take a second look at how to handle team allowances. With the help of four consultants the allowances were adjusted again 2 years later. See Sect. 2 for a full discussion (USGA, 2009).

1.2 The USGA Handicap System

The USGA handicap system is based on every player carrying an index. This index is calculated from the mean of the 10 best among the last 20 rounds of golf played, times 0.96. The last factor was introduced to favor good players with low handicaps as it dampens handicaps by a factor and not a fixed amount. It is therefore usually referred to as the "bonus for excellence". From the handicap index and some course related parameters the player receives a course handicap. This integer handicap tells the player how many strokes to deduct from the gross score to get the net score on that particular course. Every golf course ranks its 18 holes in descending difficulty order so that a player with course handicap 1 will deduct one stroke on the toughest hole, a player with course handicap 2 on the two toughest holes and so on. With a course handicap of 19 the player receives 2 strokes on the toughest hole and one on the remaining 17. A more thorough definition of the handicapping methodology is given on the USGA website (USGA, 2008a) and in the report by the first author (Siegbahn, 2009), on which this paper is based.

1.3 Team Golf

There are two types of team golf games mentioned in the rules of golf, four-some and fourball. Four-some is played with one ball per team, 2 players vs. 2 with alternating shots while fourball is played 2 vs. 2, one ball per player, where the best scores from each team are compared. Both forms can be played either as stroke play where the total strokes are counted after 18 holes or as match play where the unit of scoring is the number of holes won. Any tied holes are not counted. Other than these two types there are countless other unofficial games but this study focuses on fourball match play. Variants of the basic game include using the aggregate of the team score instead of the lowest score and a tiebreaker rule, sometimes called the "Australian" tiebreaker. This simply means that if a hole is tied by the lower scores, the higher scores of the two teams are then considered to determine a possible winner of the hole. For example, if Team A shoots a 4 and 5 on a hole while Team B shoots a 4 and 6 Team A will win the hole if the Australian tiebreaker is used, while the hole will be tied with ordinary rules.

In team golf the USGA has designed an additional punishment to players with high handicap on top of the 96% bonus for excellence. This *handicap allowance* percentage is defined for the different team games and is applied to the player course handicaps before allocating handicap strokes (USGA, 2008b).

1.4 Assumptions

For the purpose of this study, certain simplifying assumptions were made. It was assumed that golf is played on 18 identical holes, all with par 4, meaning that they were designed to require four strokes to complete. To have the players' handicap index equaling the course handicap it was also assumed that the course had a slope rating of 113 and a course rating of 72. It should be noted that these last two assumptions were not vital for this study since it is the purpose of the slope system to assure that the resulting handicap differential is independent of what course and ratings are played.

For example, let us look at two different players with handicap index 15 and 20 playing on two different courses. The better player is playing a hard course which gives him a course handicap of 18. The other player is playing an easier course which gives him a course handicap of 18 as well. If the handicap system were perfect these two players with the same course handicap, but different handicap indexes, would perform equally in the long run.

2 Previous Work

After the unwelcomed change in handicap allowances in 1976 the USGA consulted Dr. Clyne Soley, Dr. Francis Scheid, Trygve Bogevold and Dr. Richard Stroud to analyze the situation (USGA, 2009). They formed three teams, looking at the problem from different angles.

2.1 Dr. Clyne Soley

Soley used scorecards from actual team events to build a database. He came to the conclusion that was ultimately adopted by the USGA in 1978. Fourteen months of field testing showed no significant advantage to any handicap level. Table 1 shows how the USGA varied allowances before settling on the ones recommended by Soley in 1978.

Table 1 History of handicap allowances

	Pre 1976(%)	1976–1978(%)	1978–(%)
Fourball match play	100	80	100
Fourball stroke play	100	80	90

2.2 Dr. Francis Scheid

Scheid worked with a collection of 28,000 hole-by-hole rounds of 1,400 golfers. He grouped and regrouped individual players and simulated more than a million rounds trying different allowances. Though Scheid mainly looked at stroke play, his two basic findings were that no single allowance could make all teams equal and that the spread in handicap between teammates was a very significant factor. By studying average team scoring performances he selected the handicaps 5, 10, 15 and 20 and formed ten possible team setups to simulate 800 stroke play tournaments. The results from the simulation are given in Table 2. The top row and leftmost column form the teams of two players and the numbers inside show how many times out of the 800 games the combination finished in the top quarter. To hold its own, a combination would have to finish in the top quarter 200 times. From the table we conclude that the combination of two players with handicap 5 finished in the top quarter 99 times whereas a 15 and 5 did 235 times.

Table 2 Performance of different team setups. Numbers show how often out of 800 the team setup finished in the top quarter of a simulated stroke play tournament

Course handicap	5	10	15	20
5	99			
10	200	170		
15	235	234	169	
20	212	273	221	187

2.3 Trygve Bogevold and Dr. Richard Stroud

Working with the same data that Soley used the Bogevold-Stroud team focused on the effects of spread. They reported that each stroke of spread in handicap lead to an approximate advantage of one tenth of a hole in a match competition. This meant that a team of handicaps 0 and 10 would be, on average, 1 hole up on a team with handicaps 5 and 5.

3 Data Collection

This kind of study requires large amounts of data. There are two levels of detail in data collected from golf play: 18-hole data where the total strokes from 18 holes is given (also referred to as round-by-round data), and hole-by-hole data containing the score on each of the 18 holes. The latter is, of course, more detailed and therefore preferred. Unfortunately no such data was available, making the approach of this section necessary.

3.1 Data Source: Golfnet

Golfnet.com is a website offering various handicapping services to golf clubs. One of those services is handicap management. Available on the site is therefore large amounts of player round-by-round data. Roughly 25,000 Florida golfers were semi-randomly picked from the database. To get a homogeneous player base all males with a record of at least 20 rounds played on 18 hole courses, and with a handicap index between 0 and 36 were selected. This resulted in a total of 12,851 entries with 20 rounds each.

3.2 Curve Fitting

For each of the 12,851 players the handicap index, mean handicap differential and variance were calculated from their respective 20 rounds. The handicap index was calculated according to the description in (Siegbahn, 2009). The mean differential was simply the arithmetic mean $\overline{x}$ of the $N = 20$ rounds. The unbiased variance σ^2 was

$$\sigma^2 = \frac{1}{N-1}\sum_{n=1}^{N}(x_n - \overline{x})^2$$

with $N - 1$ in the denominator since the last 20 rounds constitute a sample rather than a population.

Scatter plots of all players' mean differentials and variances can be found in (Siegbahn, 2009). The pattern of the mean differentials suggests a first degree least-squares fit with the shape of

$$\text{mean differential(hcp index)} = 1.0887 \cdot \text{hcp index} + 2.1025.$$

Not surprisingly the slope of the line is slightly steeper than 1 partly because of the bonus for excellence mentioned on page 145. There is also a constant 2.1025 originating from the fact that handicap index is calculated from the mean of the best rounds, thus overestimating the player's actual skill level.

The variance dependence on handicap is not as obvious as in the case of the mean differential. A linear fit would at some point be below zero giving a forbidden negative variance. Therefore, an exponential fit seemed more appropriate with the shape of variance(hcp index) $= \alpha \cdot e^{\beta \cdot \text{hcp index}} + \gamma$. The three parameters were determined with the Matlab command `nlinfit()` giving

$$\text{variance(hcp index)} = 6.7671 \cdot e^{0.038313 \cdot \text{hcp index}} + 2.3733.$$

From here on, when using the formulas for mean and variance, the "hcp index" will be referred to as only "hcp" according to Sect. 1.4.

3.3 Hole by Hole Distributions

The purpose of going through round-by-round data was to use the properties of equidistributed random variables to determine the player hole-by-hole mean and variance. By assuming that golf is played over 18 identical par 4 holes, with x_i being the net score on hole i, the relation between hole-by-hole and round-by-round data was

$$\begin{aligned} E(x_{\text{round}}) &= E(x_1 + x_2 + \cdots + x_{18}) \\ &= E(x_1) + E(x_2) + \cdots + E(x_{18}) \\ &= 18 \cdot E(x_{\text{hole}}) \end{aligned}$$

since all holes were treated as equidistributed. The variance behaved as

$$\begin{aligned}
\mathrm{Var}(x_{\text{round}}) &= \mathrm{Var}\,(x_1 + x_2 + \cdots + x_{18}) \\
&= \textstyle\sum_{i=1}^{18} \sum_{j=1}^{18} \mathrm{Cov}(x_i, x_j) \\
&\quad \{\mathrm{Cov}(x_i, x_j) = 0 \quad \forall\, i \neq j\} \\
&= \textstyle\sum_{i=1}^{18} \mathrm{Cov}(x_i, x_i) \\
&= \mathrm{Var}(x_1) + \mathrm{Var}(x_2) + \cdots + \mathrm{Var}(x_{18}) \\
&= 18 \cdot \mathrm{Var}(x_{\text{hole}})
\end{aligned}$$

with the same conclusion as for the mean value thanks to the independently distributed hole scores. The fitting of the data together with these results suggested that the hole-by-hole distributions had the following properties

$$E(x_{\text{hole}}) = \frac{1.0887 \cdot \text{ hcp } + 2.1025}{18} \tag{1}$$

$$\mathrm{Var}(x_{\text{hole}}) = \frac{6.7671 \cdot e^{0.038313 \cdot \text{ hcp}} + 2.3733}{18}. \tag{2}$$

It remained to construct discrete hole-by-hole probability distributions. These distributions did not necessarily need to match any known distribution. It was however convenient to generate them according to some pre-designed formula since a unique distribution for each integer handicap was needed. Because the USGA handicap system (USGA, 2008a) has a feature known as Equitable Stroke Control (ESC) there is a limit of at most 9 in the recorded scores of golfers with handicaps up to 36, the probability of a score of 10 or higher was set to zero. Since a score of eagle (2 strokes under par) or better on a par 4 for an amateur golfer is unlikely, the probability for shooting a 1 or 2 was also set to zero. This left 7 unknown probabilities $p_3, \ldots, p_9$. With only 3 equations, an optimization problem was formulated to determine the probabilities as follows.

- All probabilities had to sum to 1.
- The mean result had to match Eq. 1.
- The variance had to match Eq. 2.
- All probabilities had to range between 0 and 1.
- The objective function was to minimize the squared difference between all adjacent probabilities with p_2 and p_{10} set to zero.

$$\begin{aligned}
\min\; & p_3^2 + (p_3 - p_4)^2 + (p_4 - p_5)^2 + (p_5 - p_6)^2 + \\
& (p_6 - p_7)^2 + (p_7 - p_8)^2 + (p_8 - p_9)^2 + p_9^2
\end{aligned}$$

$$\text{s.t} \sum p_i = 1$$
$$\sum p_i \cdot i = \text{mean}$$
$$\sum (i - \text{mean})^2 \cdot p_i = \text{variance}$$
$$0 \le p_i \le 1$$

The purpose of the minimization was to emphasize a smooth transition between probabilities. In short, it was a way to avoid spikes in the discrete probability density function. Figure 2 shows the distributions for handicaps 0–31. The model does not

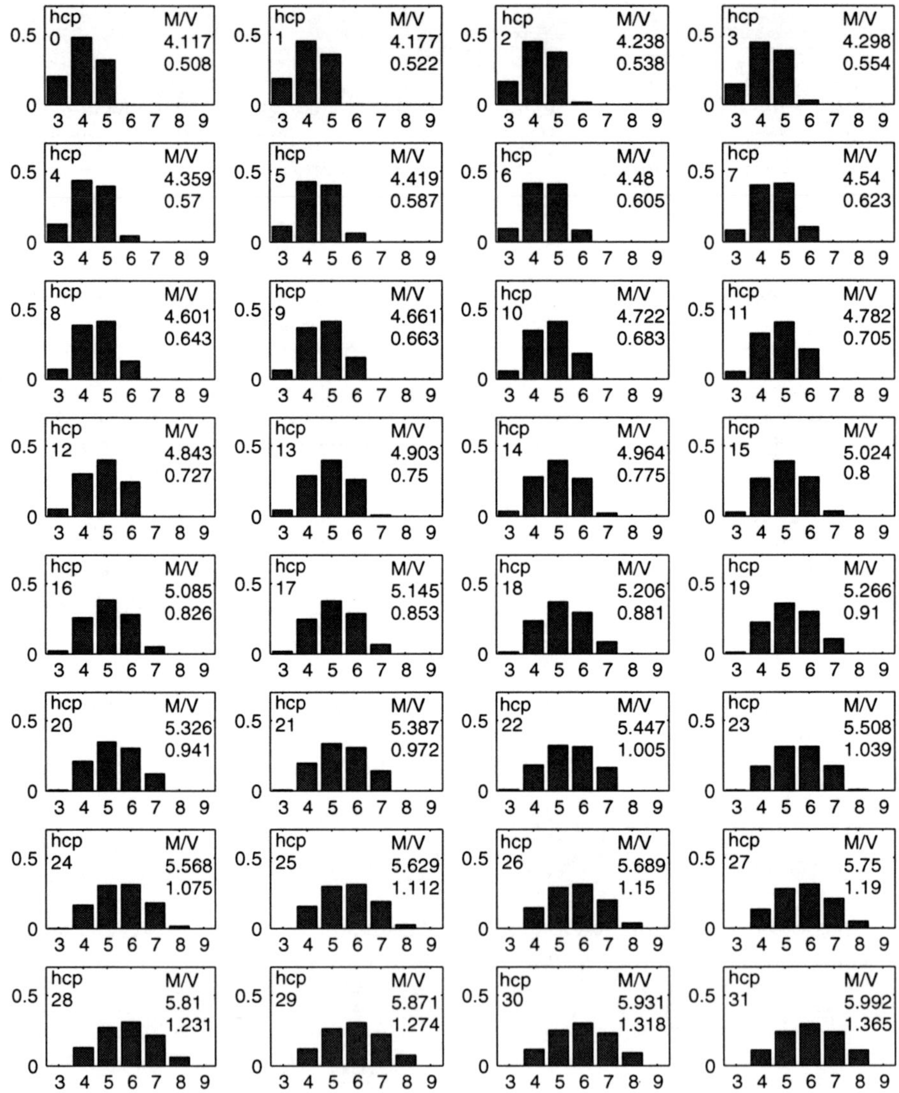

Fig. 2 Hole-by-hole distributions for players with handicap 0–31. The *bars* show the probabilities of gross scores on a par 4. M and V are mean value and variance respectively

take into account the natural skewness of player performance. True distributions tend to be positively skewed with a tail going out in the positive direction. This tail is a result of the fact that it is a lot easier to have a higher score than a lower one.

3.4 Validation

The previous sections described a method to assign probability distributions to any handicap, converting round-by-round data to the more desirable hole-by-hole data. To validate the results, a comparison was made with data published by Francis Scheid in Golf Digest (2000), see Table 3. Figure 3 shows the comparison. The means and variances coincide except for higher handicaps because the true variance is capped by the Equitable Stroke Control. Assuming that Scheid's data is good, the conclusion we draw from this section is that using our model to estimate mean and variance for any handicap is reasonable.

Table 3 Hole-by-hole data presented by Francis Scheid in Golf Digest. The numbers show gross score results from 10 rounds of 18 hole golf. Tiger's score is added for comparison but has no impact on this study

Handicap	Eagles	Birdies	Pars	Bogeys	Double bogeys	Triple bogeys	Worse
Tiger	2	48	112	17	1	0	0
0	0	27	108	40	5	0	0
5	0	10	95	60	15	0	0
10	0	6	65	80	26	3	0
15	0	3	46	83	39	7	2
20	0	1	30	76	54	15	4
25	0	1	16	66	64	25	6
30	0	1	8	44	78	36	13

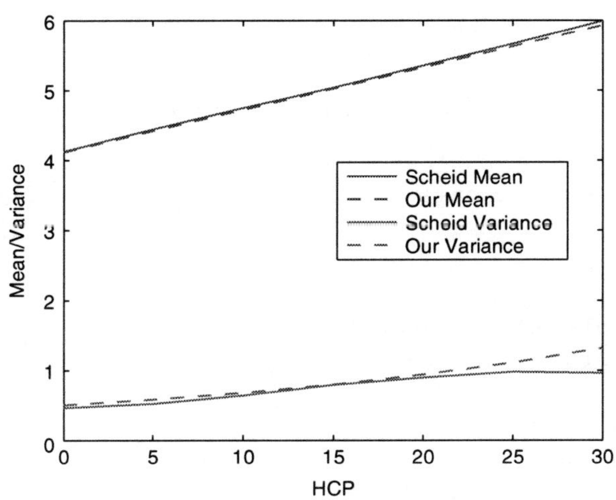

Fig. 3 Comparison of par 4 hole-by-hole data presented by Scheid vs data calculated according to our model

4 Method

With data available there were two different approaches in using it. In Sect. 2 three earlier studies were presented. They all featured an approach where actual results from score cards were used by grouping and regrouping teams to simulate matches. This study instead used data to build performance distributions of every handicap with which one could analyze all possible outcomes of the game and determine the expected results.

4.1 The Algorithm

The data collected from Golfnet gave rise to formulas for mean and variance by regression. A flowchart of the algorithm is given in (Siegbahn, 2009). The output consists of probabilities of each team winning, the chance of a draw and the *expected result*. The latter is a measure of how many holes advantage the team has. It is calculated as a regular expectation value summing the product of scores and probabilities according to

$$E(\text{result}) = \sum_{i=-18}^{+18} p_i \cdot \text{result}_i, \tag{3}$$

where $result$ is the final result of the round. Winning all holes would give +18, losing all holes would mean –18 while a score of 13–5 would yield +8. After summing all those results with the corresponding probabilities p_i we get the expected result. This measurement is a convenient way to compare how different teams perform relatively. The absolute value of the expected result is also a way to describe unfairness as the purpose of handicaps is to expect an equal game. The simulation methodology is explained in mathematical terms in (Siegbahn, 2009). Two examples of the input and output of the algorithm are given below:

Example 1:
Here we chose to input the teams manually to get the expected results.

```
Please choose an option:
   Enter teams manually                          (1)
   Enter hcps of players to find fairest teams   (2)
Pick 1 or 2: 1

How many players in Team A: 2
How many players in Team B: 2
Enter the handicap of Player 1 in Team A: 3
Enter the handicap of Player 2 in Team A: 12
Enter the handicap of Player 1 in Team B: 8
Enter the handicap of Player 2 in Team B: 10
```

```
Please choose an option:
   Regular fourball, use minimum score (1)
   Regular with Australian tiebreaker  (2)
   Aggregate, adding team scores       (3)
Pick an option from the list: 1
Probability of a tied game:          12.228%
Probability of Team A winning:       47.84%
Probability of Team B winning:       39.931%
Expected # of holes up for Team A: 0.3223
```

Example 2:
In this example we enter 4 handicaps and let the algorithm sort the teams for us. The fairest setup is presented along with corresponding results.

```
Please choose an option:
   Enter teams manually                            (1)
   Enter hcps of players to find fairest teams   (2)
Pick 1 or 2: 2
How many players in Team A? (1-4): 2
How many players in Team B? (1-4): 2
Enter the handicap of Player 1: 3
Enter the handicap of Player 2: 8
Enter the handicap of Player 3: 10
Enter the handicap of Player 4: 12
Please choose an option:
   Regular fourball, use minimum score (1)
   Regular with Australian tiebreaker  (2)
   Aggregate, adding team scores       (3)
Pick an option: 1

There are 3 ways to form teams.
Each row gives the team setup and their expected results.

3    8   |   10   12   |   0.0688
3   10   |    8   12   |   0.0786
3   12   |    8   10   |   0.3223

Column 1 to 2 form Team A
Column 3 to 4 form Team B
Column 5 gives the expected outcome for Team A

The fairest team setup would be:
Handicaps 3 & 8 playing against 10 & 12
```

```
Probability of a tied game:          12.374%
Probability of Team A winning:       44.683%
Probability of Team B winning:       42.944%
Expected # of holes up for Team A: 0.0688
```

5 Results

This section presents and tests four hypotheses all based on either intuition or previous research.

5.1 Hypotheses

- *High handicap and hence high variance improves team results.*
  Intuition suggests that high variance is an advantage since the opposite, a mutual zero spread, would be the same as playing without a partner. The conclusion can be drawn from this simplified example: consider a match where one team consists of two players with zero variance shooting par 4 on every hole. The other team's players have the very same mean around par but have an individual fifty-fifty chance to score a 3 or a 5 thereby having a nonzero variance. To win a hole the second team is only required to have one of the two players shooting a 3. This occurs on 75% of the holes since $P(3) = 1 - P(5) = 1 - 0.5^2 = 0.75$.
- *Team handicap spread improves results.*
  Dr. Francis Scheid concluded that spread in handicaps is a very significant factor. Trygve Bogevold and Dr. Richard Stroud reported that each stroke of spread in team handicap leads to an approximate advantage of one tenth of a hole in a match competition.
- *Australian tiebreaker dampens the unfairness.*
  Intuition suggests that the Australian tiebreaker rule reduces the unfairness since it dampens the positive effects of variance.
- *The fairest handicap allowance for fourball match play is 100%.*
  USGA changed to an 80% allowance in 1976–1978 but it was restored to 100% after Dr. Clyne Soley's research.

5.2 Testing of the Hypotheses

When testing the hypotheses two approaches were used:

- *Algorithm:* Running the algorithm with the discrete distributions from Fig. 2.
- *Analysis:* Approximating player distributions as continuous uniform and using probability theory.

For the second approach, though it was not realistic, it was assumed that players perform continuously uniform $X \in \mathrm{U}(a, b)$ on a hole by hole basis. On the interval $a \leq x \leq b$, with x being the net score relative par, the probability density and cumulative distribution functions were given by

$$f(x) = \frac{1}{b-a} \tag{4}$$

$$F(x) = \frac{x-a}{b-a} \tag{5}$$

with mean and variance

$$E(x) = \frac{a+b}{2} \tag{6}$$

$$\mathrm{Var}(x) = \frac{(b-a)^2}{12}. \tag{7}$$

Each hypothesis was tested by running the algorithm. Where applicable, the analysis method was used as well.

5.3 Variance

5.3.1 Algorithm

To test the first hypothesis the handicaps x for Team A were iterated from 0 to 36 while the other two players on Team B were kept at handicap 0 i.e. $[x\ x]$ vs $[0\ 0]$. The results are plotted in Fig. 4.

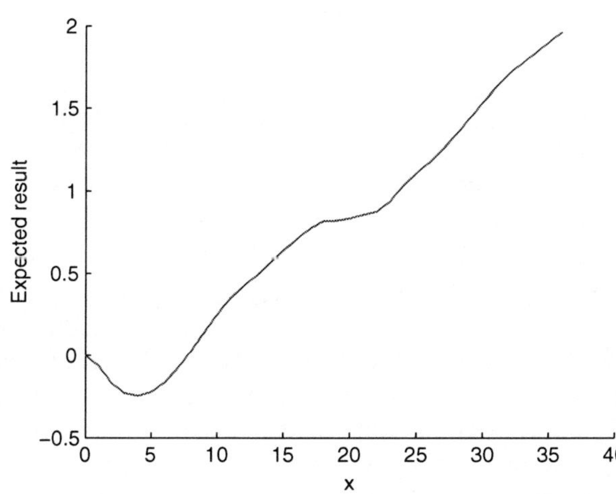

Fig. 4 Expected result in favor of Team A in a fourball match where their handicaps were iterated from 0 to 36 while the other two players of Team B remained at hcp 0. Illustrated as $[x\ x]$ vs $[0\ 0]$

We can see that the hypothesis was correct about the variance improving the results in general. The kink at low handicaps is a result of the fact that match play gives no reward for wining holes by a large margin. For low handicaps, Team A is going to overperform on a few holes so that some capacity goes to waste. Meanwhile, the opponents spread out their advantage evenly on the remaining holes. This gets repeated at handicaps around 20 for the same reason. The effect should only appear in match play but it is beyond the scope of this study to examine it for stroke play.

5.3.2 Analysis

The distribution of the minimum of two trials, the first order variable, $X_{(1)} = \min\{X_1, X_2\}$, is given by

$$f_{X_{(1)}}(x) = 2(1 - F(x))f(x) \tag{8}$$

with the expected value

$$E(X_{(1)}) = 2\int_a^b x \cdot (1 - F(x))f(x)\,dx. \tag{9}$$

Since the idea is to compare different densities with equal mean but different variance the mean is locked on 0 by setting $a = -b$. The interval then becomes $-b \leq x \leq b$ with variance $\text{Var}(X) = \frac{1}{3}b^2$. Equations 4 and 5 can be substituted in 9 to give

$$E(X_{(1)}) = -\frac{1}{3}b = -\sqrt{\frac{\text{Var}(X)}{3}}. \tag{10}$$

This result suggests that team hole scores should improve as individual variance increases, all else being equal.

To compare with the algorithm approach in Fig. 4 the distributions were approximated as continuous uniform. The mean was set to $\frac{1.0887 \cdot \text{hcp}}{18}$ according to Equation 1 with the last term removed for simplification. The variance was set to $\frac{6.7671 \cdot e^{0.038313 \cdot \text{hcp}} + 2.3733}{18}$ according to Eq. 2. Again, the games to be analyzed were $[x\ x]$ vs $[0\ 0]$. The example game [9 9] vs [0 0] is illustrated in Fig. 5. The top right graph shows the individual distributions. The top left graph shows the same distributions after adjusting for handicap strokes. The bottom graphs show the team distributions. The dashes mark the expectation values. Team A wins 72.29% of the first 9 holes while Team B wins only 69.02% of the last 9 holes. This translates into Team A being up 0.59 holes after an 18 hole round. These calculations were done for $x = 0, 1, 2, \ldots, 18$. The results are plotted in Fig. 6 and the derivation in appendix 7. Hole-by-hole results can be found in (Siegbahn, 2009).

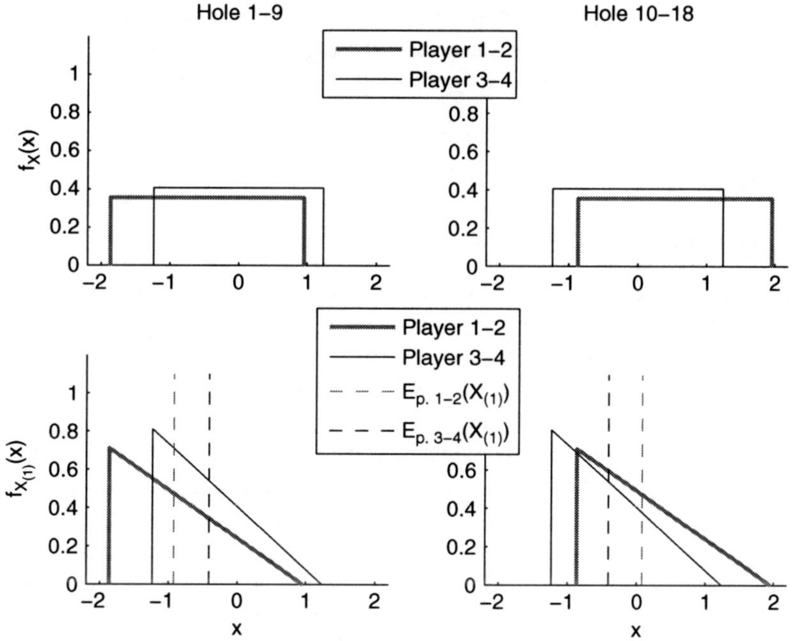

Fig. 5 *Row 1:* Players aproximated with continuous uniform distributions in the game [9 9] vs [0 0]. *Row 2:* Probability density functions and expectation value for the first order variable

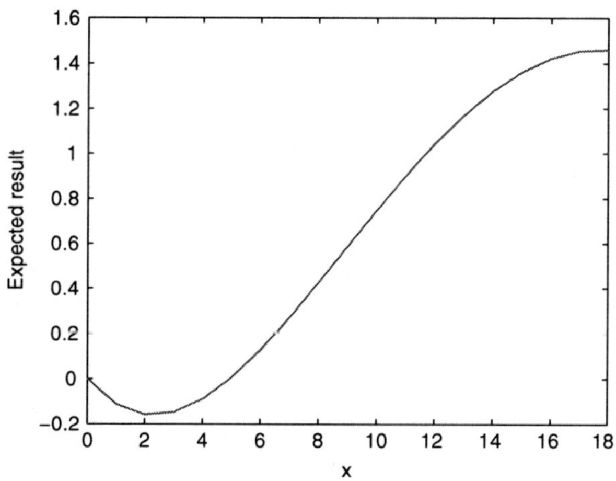

Fig. 6 Theoretical example of a [x x] vs [0 0] fourball match where the player distributions were approximated as continuous uniform with variance as in Fig. 4

The kink for low handicaps in Fig. 6 appears also when using continuous uniform distributions. This supports the analysis in the algorithm section as it shows that it is a general issue and not necessarily related to the distributions.

5.4 Spread

5.4.1 Algorithm

The second hypothesis suggests to keep three players at hcp 0 and only iterate Player 1 according to [x 0] vs [0 0]. The results are plotted in blue in Fig. 7 where one can see tendencies to the hypothesized relation until the spread reaches 10 and the result suddenly declines. All in all the blue plot somewhat matches the hypothesis but with a lower ratio and an underlying "M" shape.

To investigate this unpredicted behavior the handicap dependency of the variance was temporarily removed making it a constant for everyone. The formula for gross mean result was also altered from $4 + \frac{1.0887\text{hcp}+2.1025}{18}$ to $4 + \frac{\text{hcp}+2.1025}{18}$ to ensure that players with handicap 0, 18 or 36 would perform equally net. After doing these adjustments the result was anticipated to be a flat line with an expected result of zero no matter the handicap. Instead it was found, plotted in red, an underlying "M"-shaped behavior. Despite all players having equal variance on each hole and equal net mean after the round there was something else affecting the outcome. On three occasions the expected result was still zero as anticipated: $x = 0$, $x = 18$ and $x = 36$. What made those points special was that they left no remainder after division by 18. This meant that Player 1 got the same amount of strokes on every hole. As soon as the remainder from division was nonzero the player was going to

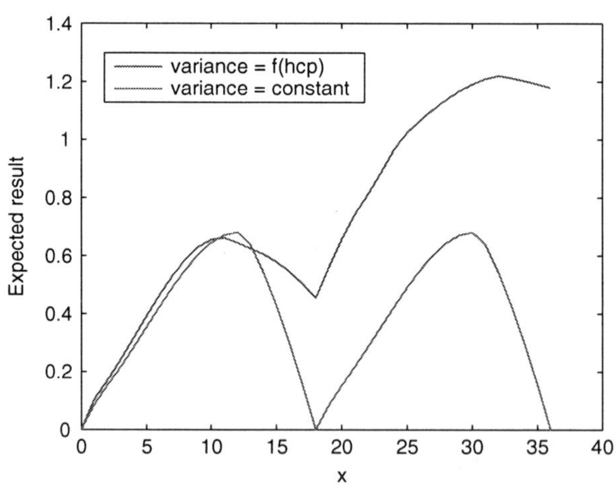

Fig. 7 Expected result in favor of Team A in a fourball match where one of their handicaps was iterated from 0 to 36 while the other three players remained at hcp 0. Illustrated as [x 0] vs [0 0]. The *red curve* shows the underlying behavior when the player variance is kept constant

receive extra strokes on some of the 18 holes. The player would see good results on holes where he got strokes but as his handicap increased his general skill and mean performance would fall meaning he would perform worse on the rest of the holes. All in all, what the player experienced is an effect coming from the discrete distribution of handicap strokes, i.e. it introduces another source of variability to the individual score. As explained earlier variance contributes to improving team results and that is why we see the "M"-shape and the kink in the blue plot.

To show the effect of spread on a more general level all possible team setups of players with handicap 0–36 were let to play against every other setup. All teams were grouped according to the spread between the two players and the mean expected result of the games was computed. The resulting plot in Fig. 8 shows that spread is a significant factor. The two maxima are not located far from the peaks in Fig. 7.

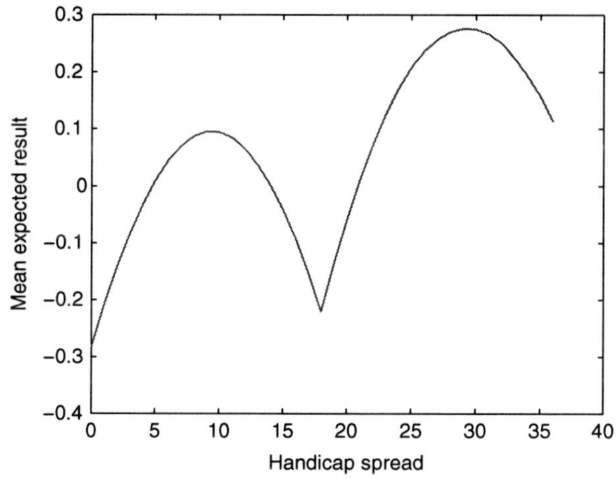

Fig. 8 Letting all possible team setups of players with handicap 0–36 play against each other, grouping by spread and taking the mean expected result gives a general view on how spread affects the performance

5.4.2 Analysis

Again the theoretical approach involved approximating the distributions as continuous uniform. The situation from Fig. 7 was picked where the altering player was at handicap 9, meaning that the game was played [9 0] vs [0 0]. The formula for gross mean result was altered from $4 + \frac{1.0887\text{hcp}+2.1025}{18}$ to $4 + \frac{\text{hcp}}{18}$ similar to the methodology in the algorithm section but with the last term removed for simplification. Variance was locked on $\frac{1}{3}$ so that Player 2–4 performed a net score relative par of $U(-1, 1)$ while player 1's distribution was shifted $\frac{9}{18}$ to the right because of his lesser skill. On hole 1–9 however, he got one stroke from handicap and therefore also shifted his distribution to the left by one stroke. The first two plots in Fig. 9 illustrate

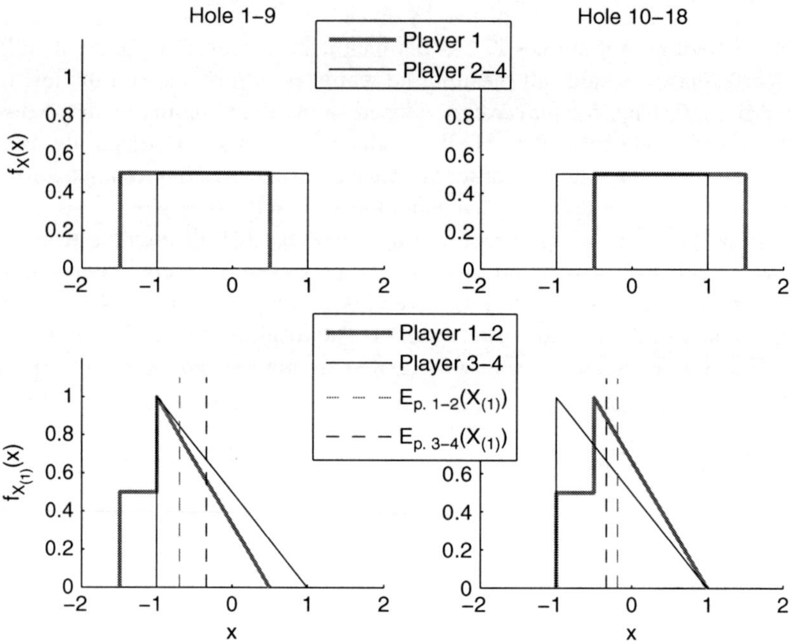

Fig. 9 *Row 1:* Players approximated with continuous uniform distributions in the game [9 0] vs [0 0]. *Row 2:* Probability density functions and expectation value for the first order variable

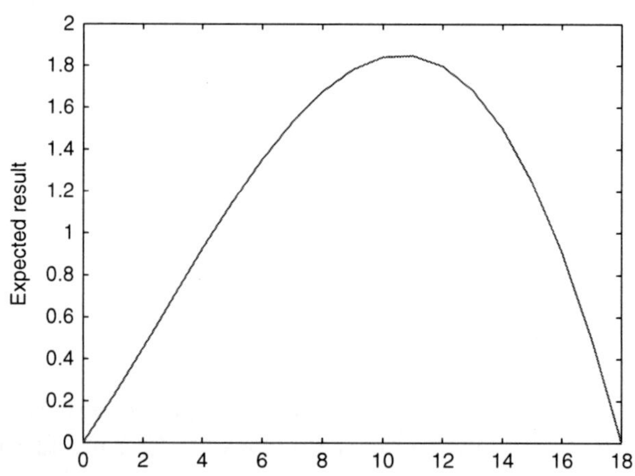

Fig. 10 Theoretical example of a [x 0] vs [0 0] fourball match where the player distributions are approximated as continuous uniform with equal variance

this. The last two plots show the density of the first order variables. The expectation values are marked with dashes and suggest that Team A has a bigger advantage on hole 1–9 than Team B has on hole 10–18. In this example, when playing [9 0] vs [0 0] with the theoretical uniform distributions, Team A has a 69.53% chance of winning each of the nine first holes while the chance of losing on the remaining holes is only 59.64%. This summarizes to Team A being up 1.78 holes after an 18 hole round. This exercise was repeated for the remaining games in [x 0] vs [0 0]. The results are plotted in Fig. 10 and the derivation in appendix 8.2. Hole-by-hole results can be found in (Siegbahn, 2009).

With $x = 0$ or $x = 18$ the game is perfectly equal but any other handicap will give rise to an advantage for Team A. This agrees well with Figs. 7 and 8. It is therefore not true that handicap spread between partners is good in all situations. What should be sought after is getting strokes on some holes and letting your partner excel on the other holes. This effect will from here on be referred to as *stroke allocation variance*.

5.5 Australian Tiebreaker

Because of the discrete nature of the Australian tiebreaker it would not be valid to approximate the distributions as continuous uniform. This left the algorithm approach only.

5.5.1 Algorithm

To try the impact of the Australian tiebreaker the same tests as before were run but with the rule activated. The variance was tested in the game [x x] vs [0 0] in Fig. 11

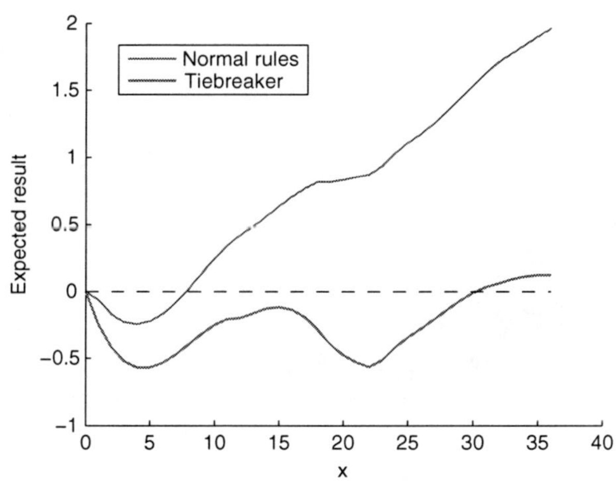

Fig. 11 [x x] vs [0 0] with normal rules plotted in *blue*. Australian tiebreaker rules plotted in *red*

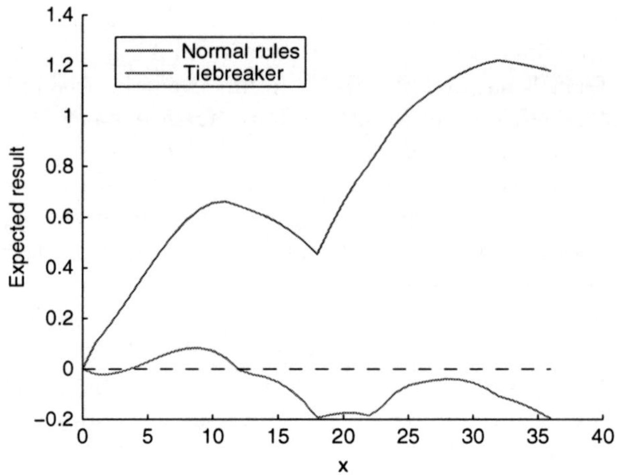

Fig. 12 [x 0] vs [0 0] with normal rules plotted in *blue*. Australian tiebreaker rules plotted in *red*

and handicap spread in [x 0] vs [0 0] in Fig. 12. It is clear that the tiebreaker rule keeps the expected results at a more reasonable level.

Another interesting test involved calculating expected result for every possible team setup composed by players in the handicap range 0–36. The results were collected in a histogram in Fig. 13 to show the total effect of the Australian tiebreaker.

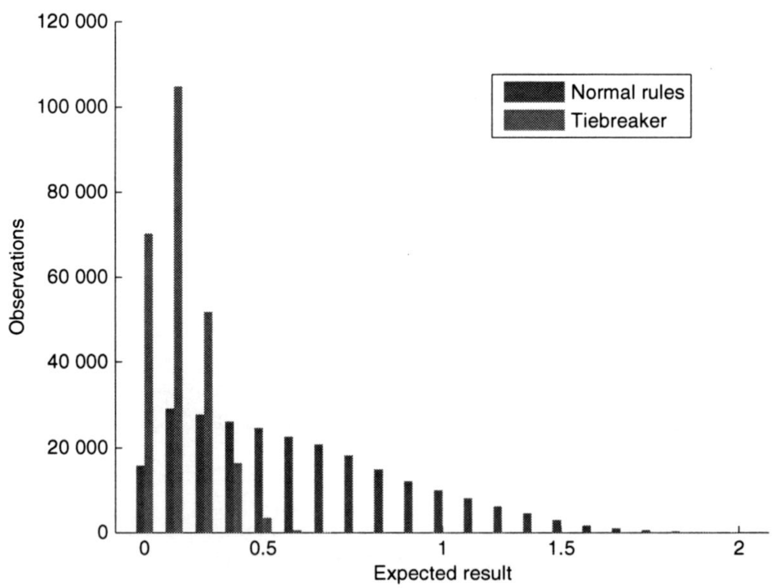

Fig. 13 Working with handicaps 0–36 there are 247,456 unique matchups. This histogram shows how the observed expectations from normal rules and tiebreaking rules are distributed

The absolute values of the expected results of all games were divided into bins. The mean for the normal games was at 0.517 and at 0.114 with tiebreaker. This suggests that, on average, the Australian tiebreaker would lower the unfairness of games played with players 0–36 by more than 70%. The most unfair game under normal rules had an expected result close to 2 while with the tiebreaker rules it never exceeded 0.5.

To make sure that this observation holds for other player bases than 0–36 the expected result was measured for teams made up of handicaps 0 to x. See Fig. 14.

The plot tells us that, under normal rules, the broader the player base the more unfairness is anticipated. Tiebreaking rules, on the other hand, show a more stable tendency. For comparison some example results are given in Table 4 where one can see the impact of the Australian tiebreaker on a more detailed level.

Table 4 Expected results for some games. Expectations are in favor Team A

Team A		Team B		Expected result	Expected result
Player 1	Player 2	Player 3	Player 4	normal rules	Tiebreaker
3	3	11	18	–1.032	–0.117
1	1	12	18	–0.982	–0.001
0	10	5	5	0.714	0.309
2	17	2	6	0.288	–0.073
3	12	8	10	0.322	0.177
3	8	10	12	0.069	0.386
2	4	7	9	0.000	0.356
0	10	7	13	0.000	0.219

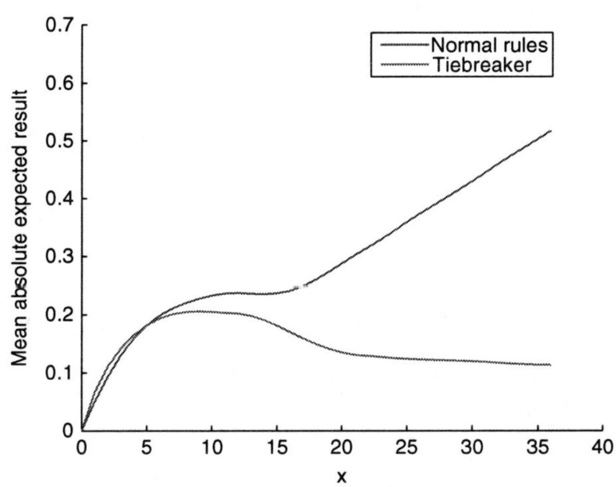

Fig. 14 This plot shows the mean of the absolute expected results as a function of player base. With $x = 0$ we only look at the game [0 0] vs [0 0]. With $x = 36$ all possible matchups consisting of players handicapped 0–36 were computed

The damping effect witnessed in Figs. 11 and 12 also seemed to affect the fairness in general. Bear in mind that some games are actually less fair with tiebreaking rules. An example would be the game [5 5] vs [0 0] in Fig. 11 where the red plot deviates more from equilibrium than the blue plot in favor of Team B. The last three rows in Table 4 show this as well.

5.6 Handicap Allowance

Because handicap allowances require realistic distributions to be meaningful the analysis with uniform distributions was not pursued.

5.6.1 Algorithm

When allowances are introduced to the game, course handicaps are damped down and then rounded to the nearest integer before stroke allocation. For example, in Fig. 7, the goal would be to flatten out the blue curve by penalizing higher handicaps. Different allowances were experimented with and the results are plotted in Fig. 15.

At 100% there is no allowance and the curve is unchanged. At 95% two kinks lower the result. In [10 0] vs [0 0] and 95% allowance the 10 is lowered to 9.5 and then rounded back to 10. In [11 0] vs [0 0] and 95% allowance the 11 is lowered to 10.45 and then rounded to 10 thus giving the team one whole stroke less. This discrete behavior is the reason for the kinks. The figure shows that the game [36 0] vs [0 0] is fair at allowance 85%. On the other hand [18 0] vs [0 0] is more fair with 90% and [9 0] vs [0 0] with 85% again. Another problem is that although one particular game gets fairer with allowance other games do the contrary. We learned earlier that stroke allocation variance has a big impact on the results. Sometimes

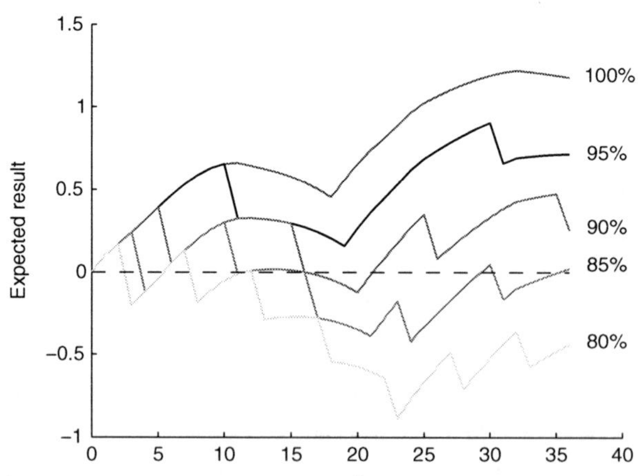

Fig. 15 Different allowances, given in percent, tried on the game [x 0] vs [0 0]

this contribution is larger than the one from the player variance. A simple example is the game [9 0] vs [18 0] favoring Team A with 0.1846 holes. With allowance the results would be even more unfair since it would punish Team B more. This suggests that in some games the allowance needs to be higher than 100% instead making it a bonus for high handicaps. It is therefore impossible to introduce one allowance that makes all games more fair. On the other hand one could try to laborate with an allowance that makes games in general more fair, just like it was done for the Australian tiebreaker in Fig. 13. The problem is that allowances are more sensitive to handicaps than what tiebreaking rules are. We examined the 18,145 matchups from teams consisting of players 0–18 and found in Fig. 16 that a 96.5% allowance was fairest. In reality what 96.5% allowance means is that players with handicap between 15 and 42 will get one stroke less since $0.965 \cdot 14$ rounds to 14 while $0.965 \cdot 15$ rounds to 14 too. If we, for instance, examine players with hcp 0–19 or 0–36 instead of 0–18 some other allowance might be fairer, but calculations of that magnitude would require weeks to complete.

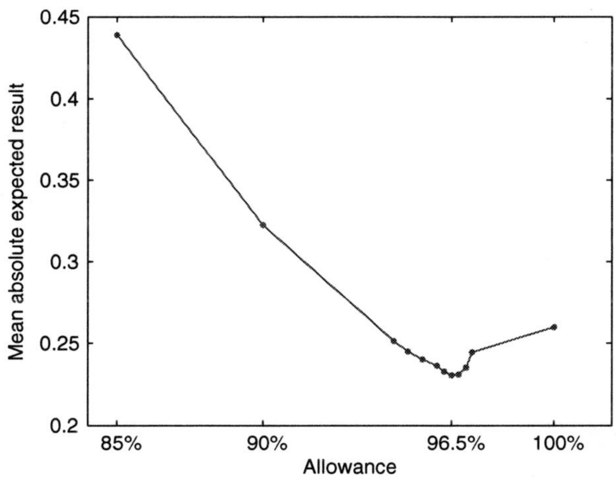

Fig. 16 Working with handicaps 0–18 there are 18,145 unique matchups. This plot shows the mean of the absolute values of all expected results as a function of the allowance

6 Summary and Conclusion

6.1 Variance

The contribution to unfairness from player variance was the issue that brought this study to life. The problem is not fresh news, it has rather been a known issue since at least 1976 when multiball allowances were introduced to counter the effect. The conclusion we draw from the results is that variance plays a significant role in the unfairness. We were also able to prove the theory behind the effect by order statistics. The previous study by Dr. Francis Scheid supports our results when Table 2 shows an advantage for high handicap teams. Their results covered stroke play but the general idea should be applicable to match play as well.

6.2 Spread

The effect of spread is not mentioned or dealt with in the USGA literature. Still, we manage to show that pairing with the right partner for stroke allocation variance can be more important than regular variance. In previous studies spread in itself was considered an advantage. We have managed to prove that it is rather a team spread of around 11 or 29 that should be sought after, while a spread of 18 is the same as no spread at all when individual player variance is ignored. Our conclusion is again backed up by results by Dr. Francis Scheid in Table 2 where the best performing teams are closest to spread 11 (5 & 15 and 10 & 20).

6.3 Australian Tiebreaker

The Australian tiebreaker rule seems like a very promising way to cope with unfairness in fourball. We showed that, in general, for handicaps 0–36, it reduces the unfairness by more than 70%. It is also a very simple and intuitive change of the rules. The only downside, one could argue, is that it might go against the reason to play fourball in the first place. Fourball is a popular game because it encourages variance. People like to have a partner that can cover for the occasional bad scores. The Australian tiebreaker lowers this insurance as tied holes are very common.

6.4 Handicap Allowance

Our results suggest that if a fourball tournament were held with all teams of handicaps 0–18 represented, an allowance of 96.5% would lead to fair games in general. The problem is that 96.5% penalizes players with handicap 15–18 by one stroke and thereby makes handicap 14 very desirable. This illustrates the potential problem with allowances. As they are designed today, the two-step rounding of the handicap might give rise to an unwanted threshold effect when course handicaps are rounded to nearest integer, damped by allowance and then rounded again.

7 Directions for Future Research and Improvements

In Sect. 1.4 some assumptions were made to simplify the calculations. For example, all 18 holes were treated as par 4 and identical in difficulty. If more detailed data had been available one could have built distributions for par 3, par 4 and par 5 separately.

Since par 3 holes usually tend to be rated as easier the stroke allocation would have been different. Another way to increase the significance of the results would be to simply collect more player data. The construction of the distributions is also something that could be improved as they do not take into account the skewness. This report has considered match play only. It would be interesting to study the difference between match and stroke play as team distribution skewness probably would not have any impact on stroke play.

Appendix

Derivation of Fig. 6

The games [hcp hcp] vs [0 0] were analyzed for $0 \leq \text{hcp} \leq 18$ with variance set to

$$\text{Var(hcp)} = \frac{6.7671 \cdot e^{0.038313 \cdot \text{hcp}} + 2.3733}{18} \tag{11}$$

according to Eq. 2. The mean was set to

$$E(\text{hcp}) = \frac{1.0887 \cdot \text{hcp}}{18} \tag{12}$$

according to Eq. 1 with the last term removed for simplification. The a and b parameters in the uniform probability distributions $\text{U}(a(\text{hcp}), b(\text{hcp}))$ were thereby given by solving Eqs. 6 and 7.

$$a(\text{hcp}) = E(\text{hcp}) - \sqrt{3 \cdot \text{Var(hcp)}} \tag{13}$$

$$b(\text{hcp}) = E(\text{hcp}) + \sqrt{3 \cdot \text{Var(hcp)}} \tag{14}$$

The team distributions were the first order statistics of the individual distributions. They can be seen graphically in the bottom row of Fig. 5 with triangle shape. The left plot represents the holes where Team A got strokes from handicap, thus increasing their chances. The right plot represents the remaining holes where Team A got no strokes but still suffered from being weaker players, thus lowering their chances. The distribution for holes without strokes was

$$f_X(x, \text{hcp}) = \begin{cases} \frac{2b}{(b-a)^2} - \frac{2}{(b-a)^2}x & \text{if } a \leq x \leq b \\ 0 & \text{else} \end{cases} \tag{15}$$

Team A's probability of winning the holes where no strokes were given was

$$\begin{aligned} &p(\text{hcp}) \\ &= P(X(\text{hcp}) < X(0)) \\ &= \int_{a(\text{hcp})}^{b(0)} f_X(x,\text{hcp}) \left(\int_x^{b(0)} f_X(y,0)dy \right) dx \\ &= \frac{1}{b(0)(b(\text{hcp}) - a(\text{hcp}))^2} \\ &\cdot \left[\frac{b(0)b(\text{hcp})}{2}x - \frac{b(0)+2b(\text{hcp})}{4}x^2 + \frac{b(\text{hcp})+2b(0)}{6b(0)}x^3 - \frac{1}{8b(0)}x^4 \right]_{b(\text{hcp})}^{b(0)} . \end{aligned} \tag{16}$$

On the holes where strokes were given the distribution was shifted to the left by one stroke.

$$\hat{a}(\text{hcp}) = a(\text{hcp}) - 1 \tag{17}$$

$$\hat{b}(\text{hcp}) = b(\text{hcp}) - 1 \tag{18}$$

The distribution for holes with strokes was

$$f_{\hat{X}}(x,\text{hcp}) = \begin{cases} \frac{2\hat{b}}{(\hat{b}-\hat{a})^2} - \frac{2}{(\hat{b}-\hat{a})^2}x & \text{if } \hat{a} \le x \le \hat{b} \\ 0 & \text{else} \end{cases} \tag{19}$$

The probability of Team A winning the holes where they had the advantage of handicap strokes was

$$\begin{aligned} \hat{p}(\text{hcp}) &= P(\hat{X}(\text{hcp}) < X(0)) \\ &= \int_{\hat{a}(\text{hcp})}^{a(0)} f_{\hat{X}}(x,\text{hcp})dx + \int_{a(0)}^{\min(b(0),\hat{b}(\text{hcp}))} f_{\hat{X}}(x,\text{hcp}) \left(\int_x^{b(0)} f_X(y,0)dy \right) dx \\ &= \frac{2}{(\hat{b}(\text{hcp}) - \hat{a}(\text{hcp}))^2} \left(-\hat{b}(\text{hcp})b(0) - \frac{b(0)^2}{2} - \hat{b}(\text{hcp})\hat{a}(\text{hcp}) + \frac{\hat{a}(\text{hcp})^2}{2} \right) \\ &+ \frac{1}{b(0)(\hat{b}(\text{hcp}) - \hat{a}(\text{hcp}))^2} \cdot \left[\frac{b(0)\hat{b}(\text{hcp})}{2}x - \frac{b(0)+2\hat{b}(\text{hcp})}{4}x^2 \right. \\ &\left. + \frac{\hat{b}(\text{hcp})+2b(0)}{6b(0)}x^3 - \frac{1}{8b(0)}x^4 \right]_{-b(0)}^{\min(b(0),\hat{b}(\text{hcp}))} \end{aligned}$$

where $a(0) = -b(0)$ was used for simplification.

Derivation of Fig. 10

The games [hcp 0] vs [0 0] were analyzed for $0 \le \text{hcp} \le 18$ where the probability distribution of Team B on all 18 holes was the first order statistic of the two $Y \in$ U$(-1, 1)$ distributions. The density function was plotted in Fig. 9 with a shape of a triangle. The distribution was thus

$$f_{Y_{(1)}}(y) = \begin{cases} 0.5 - 0.5y & \text{if } -1 \le y \le 1 \\ 0 & \text{else} \end{cases} \tag{20}$$

Next to this distribution are the red plots that give us the distributions of Team A. They are both comprised of one rectangular and one triangular part. The left plot represents the holes where Player 1 gets strokes from handicap, thus increasing Team A's chances. The right plot represents the remaining holes where Player 1 gets no strokes but still suffers from being a weaker player, thus lowering Team A's chances. These two distributions are given below as $f_{\hat{X}}(x)$ and $f_X(x)$ respectively. The variable a is the rightward shift in Player 1's distribution. It is defined as $a = \frac{\text{hcp}}{18}$ with $0 \le \text{hcp} \le 18$ and hence $0 \le a \le 1$.

$$f_{\hat{X}}(x) = \begin{cases} 0.5 & \text{if } a - 2 \le x \le -1 \\ \frac{a}{a+1} - \frac{1}{a+1}x & \text{if } -1 \le x \le a \\ 0 & \text{else} \end{cases} \tag{21}$$

$$f_X(x) = \begin{cases} 0.5 & \text{if } -1 \le x \le a - 1 \\ \frac{1}{2-a} - \frac{1}{2-a}x & \text{if } a - 1 \le x \le 1 \\ 0 & \text{else} \end{cases} \tag{22}$$

The probability of Team A winning the holes where they had the advantage of handicap strokes was

$$\begin{aligned} \hat{p}(a) = P(\hat{X} < Y_{(1)}) &= \int_{a-2}^{a} f_{\hat{X}}(x) \left(\int_x^1 f_{Y_{(1)}}(y) dy \right) dx \\ &= \frac{1}{48(a+1)} \left(17 + 28a + 6a^2 - 4a^3 + a^4 \right). \end{aligned} \tag{23}$$

Similarly, the probability of winning the holes where no strokes were given was

$$\begin{aligned} p(a) = &P(X < Y_{(1)}) = \int_{-1}^{1} f_X(x) \left(\int_x^1 f_{Y_{(1)}}(y) dy \right) dx \\ &= \frac{1}{16(2-a)} \left(16 - 32a + 24a^2 - 8a^3 + a^4 \right). \end{aligned} \tag{24}$$

The simple test that $\hat{p}(1) = p(0) = 0.5$ verified that [0 0] vs [18 0] was an even game.

References

Golf Digest. (2000, May). *Your score vs. Tiger's, 51*(5), 33. ISBN 0017176X.

Pope of Slope. (2008). *History of USGA handicap procedures in the United States up to 1979.* Retrieved June 14, 2008, from http://www.popeofslope.com/history/history.html

Scheid, F. (1978). *The search for the perfect handicap.* Retrieved June 14, 2008, from http://portal.acm.org/citation.cfm?id=807622

Siegbahn, P. (2009). *A study of fairness in team golf competition.* Retrieved January 26, 2008, from http://siegbahn.com/patrick/files/pdf/thesis golf.pdf

USGA. (2008a). *Handicap manual.* Retrieved June 14, 2008, from http://www.usga.org/playing/handicaps/manual/manual.html

USGA. (2008b). *The rules of golf.* Retrieved June 14, 2008, from http://www.usga.org/playing/rules/books/rules.html

USGA. (2009). *What the new multi-ball allowances mean to you.* Retrieved October 4, 2009, from http://www.usga.org/handicapping/articles resources/What-the-New-Multi-Ball-Allowances-Mean-to-You(2)/

Can Subsidies Help Buy Success? Revenue Sharing in English Football

Rob Simmons and Ian Walker

Abstract This paper investigates the impact of redistribution of revenues from large successful organizations to small unsuccessful ones. In particular, we ask whether such redistribution actually makes the performance of these organizations closer. Such redistributive mechanisms are common in sports where the motivation is to ensure that competitions are closely run in the sense that the outcomes become hard to predict. In some cases this is achieved through regulation, in others through handicapping, and in the case of English football and many other sports financial redistribution is employed. We aim to identify the effect of such redistribution by looking at how exogenous variation in financial resources impact on success of the field. This is achieved by exploiting the random assignment of teams to games in the FA Cup knockout competition. We use this random assignment to investigate the extent to which the resulting variation in revenues impact on the performances of clubs in the League competition where all teams play each other (twice) each season.

1 Introduction

In professional sports leagues it is often claimed that a competitive equilibrium might result in too little competitive balance, in the sense that strong teams dominate weak teams "too much" and more even contests might be more attractive to fans and audiences. Consequently, sports leagues have often adopted forms of revenue-sharing in an attempt to redress perceived lack of competitive balance. The question of effectiveness of revenue sharing in sports leagues has occupied the attention of sports economists since the seminal contribution of Rottenberg (1956). According the invariance proposition, itself an application of the Coase theorem, redistribution of revenues in team sports cannot affect competitive balance in a sports league. For a two team league, with profit maximizing clubs, Fort and Quirk (1995) showed that gate revenue sharing displaces the marginal revenue curves of each team with equal

R. Simmons (✉)
Lancaster University, Lancaster, UK
e-mail: r.simmons@lancaster.ac.uk

S. Butenko et al. (eds.), *Optimal Strategies in Sports Economics and Management*,
DOI 10.1007/978-3-642-13205-6_9,

shifts so that competitive balance remains unchanged. This conclusion was questioned by Szymanski and Késenne (2004). Using a contest game-theoretic model, in which two players (teams) choose levels of talent so as to maximize profits given the choices of rivals, they obtain the result that, comparing before and after Nash equilibria, gate revenue sharing will worsen competitive balance. The intuition behind this result is that gate revenue sharing reduces the incentives for small teams to invest in player talent, relative to large teams. Small teams gain more from the success of large teams than big teams gain from the success of small teams. Then small teams reduce investment in player talent relative to large teams. Inequality in competitive balance is then exacerbated.

Gate revenue sharing and sharing of broadcast revenues are prevalent in North American sports. In American football (National Football League), 40% of gate revenues go to the away team. In Major League Baseball, the league collects a fixed share of all net local revenues and redistributes these revenues in a progressive manner such that teams with lower revenues receive a greater share of the pool of money that is redistributed (Maxcy, 2009). As Maxcy points out, this arrangement could lead to the unintended and perverse result, as identified by Szymanski and Késenne (2004), that the League subsidy reduces the incentive to win for lower income teams. This implies that such teams are more likely to engage in player trades with better talent moving away from small-revenue teams to large-revenue teams. Maxcy's empirical analysis of player trades in baseball supports this prediction.

In contrast, if teams have the alternative objective of win maximization subject to a break-even or maximum loss constraint then Késenne (2000, 2007) shows that gate revenue sharing will succeed in improving competitive balance, whether the league consists of just two teams as in Fort and Quirk (1995) or many teams. Hence, theory has the capability to offer predictions that revenue sharing can improve competitive balance, worsen competitive balance or leave competitive balance unaffected. These contrasting predictions turn fundamentally on assumptions surrounding the objectives of teams and on elasticity of talent supply to wage rates.

In European football, there appears to be a general movement away from revenue-sharing. In the English Football League, gate revenue sharing was practised until 1983, when it was formally abandoned (Dobson & Goddard, 2001). Broadcast revenues are shared in European football leagues to some extent, but lower divisions receive low levels of television coverage and there is no mechanism to redistribute revenues from teams in the top division to teams in lower divisions. Over time in English football, club revenues have shown increased dispersion, both between divisions and within divisions (Dobson & Goddard, 2004).

Empirical evidence on the effects of gate revenue sharing on league outcomes is not conclusive. Szymanski (2003) and Borland and Macdonald (2003) note the lack of consensus in attendance demand estimates on impacts of outcome uncertainty on attendance demand, either at the level of individual matches or in terms of seasonal outcomes. As with the theory of competitive balance, studies can be cited that show positive effects of outcome uncertainty on attendance demand, zero effects and negative effects. Recently, Buraimo and Simmons (2009) have shown for Spanish football that gate attendances respond negatively to increased match outcome uncertainty, as measured by match win probabilities extracted from ex ante betting odds. In contrast, broadcast audiences, measured by viewing figures, respond positively to increased match outcome uncertainty, controlling for quality of the competing teams.

The present paper offers a new perspective on the question of effectiveness of revenue sharing on league outcomes. We examine English football and exploit the random assignment property of the FA Cup knock out competition. In this competition, teams compete against each other with opponents selected by random draw. There is no seeding apart from a stipulation that teams in the top two divisions enter at a later stage than teams in the bottom two divisions. Since the draw is random, lucky teams will meet larger opponents and this offers a revenue-generating opportunity. Revenues from a Cup draw against a large opponent can be recycled into team strength so as to compete more effectively in the League competition.

A paper by Szymanski (2001) also uses data from the English FA Cup and addresses the question of relationship between competitive balance in a league and attendances. Szymanski observed that attendances at FA Cup matches for teams drawn from the same division were less, on average, than attendances at the corresponding League matches between the same pairs of teams. Szymanski interpreted this result as due to increased interdivisional competitive imbalance between league teams. According to Szymanski, the growth of income inequality between divisions has reduced interest in inter-divisional Cup matches while intra-divisional League matches are more interesting to fans than comparable Cup matches.

We pose a very specific question. If revenue sharing helps recipient teams improve their prospects, then teams drawn randomly against large opponents in the FA Cup should generate greater revenues than their divisional rivals who do not benefit from a lucrative Cup draw. If the extra resources from augmented revenues can then be capitalized into improved League performance then we can take this as evidence that revenue sharing does indeed promote competitive balance.

Teams in the bottom two divisions of the English Football League are rarely televised and there is no mechanism to redistribute television broadcast revenues to clubs at this level. The lack of shared gate or broadcast revenues then makes the FA Cup competition particularly important for teams in the bottom two divisions. This is because revenues from progress through the FA Cup are of two types: a prize money component based on the round in the tournament that is reached and sharing of revenues from gate receipts and from broadcasting of Cup fixtures. Unlike the League competition, gate revenues are shared by the competing teams in all Cup matches. Hence, the FA Cup combines the twin features of random (unseeded) draw and gate revenue sharing. Below, we shall empirically estimate the impacts of FA Cup progress through the rounds on team revenues and team performances and pose the fundamental question of whether revenue sharing through the FA Cup competition has beneficial impacts on team revenues and team performances. The impact of the FA Cup random draw will be distinguished in our empirical model from the impact of success (momentum) gained by successful progress through the competition.

The FA Premier League and the Football League together currently comprise 92 teams with 20 in the Premier League and 24 in each of the three Football league Divisions. We shall refer to these divisions as tier 1 (Premier League), tier 2 (currently called Football League Championship), tier 3 (League One) and tier 4 (League Two). Prior to 1992, these divisions would have been referred to simply as Divisions 1 to 4. The FA Premier League was formed in the 1992/1993 season, principally in order for top division clubs to capture a larger share of broadcast revenues offered by the new satellite TV company, BSkyB. Mobility of teams between divisions is obtained through promotion and relegation with three teams demoted each season from tiers 1–3 and four from tier 3, to be replaced by promoted teams from the tier below. A relatively recent innovation is the playoff competition for the third (or fourth for tier 4) promotion place. Two teams are promoted automatically from tiers 1 and 2, while teams placed third through sixth compete in a knock-out competition to secure one promotion place. This device was intended to raise audience interest at the end of the season, as more games would involve playoff contention. The promotion and relegation system contrasts sharply with the self-contained "closed" North American league structures (Noll, 2002). As Rosen and Sanderson (2001) succinctly put it, the promotion and relegation system "punishes failure". In particular, it punishes failure to invest in team-building.

The remainder of the paper is structured as follows. Section 2 explains the role of the English FA Cup in terms of its impacts on team revenues, effort and performance. Section 3 outlines a model for estimation and explains our data set. Section 4 offers empirical results while Sect. 5 concludes.

2 The FA Cup

The Football Association Cup began in the season 1871/1872 with just 15 teams competing. In 2009/2010, a record number of 786 teams took part with qualifying rounds featuring amateur teams beginning in August. As rounds progress, sets of higher status teams are added until the bottom two divisions join the First Round in November, to be followed by teams from the top two divisions in the Third Round in January. This features 64 teams, with 44 from the top two divisions by right. The remaining 20 places are filled by a mixture of lower division and non-league teams. Currently, a tied Cup game leads to a replay at the away team's ground. It the replay is also tied after extra time then the tie is settled by a penalty shoot-out. Occasionally, a small team might cede home advantage to a bigger team so as to capture extra revenue, but this is rarely practised by lower division teams. After four more knock-out rounds, the FA Cup Final is contested at a neutral venue, usually Wembley Stadium, in May. This is a glamorous and prestigious occasion, regarded as a showcase for English football and the Final is broadcast live worldwide.

Recently, media commentators have questioned whether the FA Cup has lost some of its glamour and appeal. In the 1999/2000 season, Manchester United withdrew its entry to the FA Cup, and competed in a new FIFA international club tournament instead (Dobson & Goddard, 2001). It has been observed that many Premier League clubs field weakened teams, consisting of fringe and reserve players, in earlier rounds of the FA Cup. Their argument is that they need to rest the best players for the Premier League competition. Implicit in this lack of effort to win Cup matches is the economic view that revenues from retention of Premier League status, by avoiding demotion, are much greater than revenues from progress in the FA Cup. Indeed, relegation from the Premier League to the Football League Championship is estimated to cost the unfortunate club around £60m in lost revenue, even though some broadcast revenues are distributed to relegated teams for a 2 year period following demotion.

In contrast, the opportunity for a small, lower division team presented by a random draw against a high-profile Premier League team is still considerable. Acts of "giant killing" where low level teams actually beat big teams in the FA Cup are enshrined in history and add to the romantic appeal of the competition. Such exploits are remembered (and replayed on TV) for many decades after they occur. In immediate terms, the random draw against a large club can offer considerable revenue-generating opportunities as the following example illustrates:

Colchester (tier 3) away to Chelsea (tier 1) in season 2005/2006, round 5.

Colchester beat two teams from tier 3 to reach round 5. In round 5, Colchester received £400,000 from their share of gate revenue from a crowd of 41,810 at Chelsea- Colchester's ground capacity is 7,000. Broadcast revenue from this game was £265,000. This one game raised £750,000 revenue for Colchester and their total revenues from the Cup competition came to £1m. At the end of the season, Colchester finished 2nd in their division and hence won promotion to the Championship (tier 2). This example shows that lower division teams can benefit both from progress through the rounds of the FA Cup, and also from being randomly drawn against bigger teams. Of course, the lower the divisional status of a team, the smaller is the probability of a team reaching later rounds such as round 5 or beyond. In fact, no tier 4 team has reached round 5 since the 1985/1986 season. Table 1 shows prize money and potential broadcast revenues by round of FA Cup

Table 1 Revenues per club in FA Cup 2006/2007

Round	Prize	TV income
3	40,000	150,000
4	60,000	150,000
5	120,000	265,000
6	300,000	265,000
Semi final	900,000	
Final	1,000,000	

Revenues from ticket sales of a particular Cup match, net of VAT and eligible costs, are divided as: 45% per team (home and away) 10% to a central pool for later distribution to football associations and clubs.

assessed at the 2006/2007 season. The introduction of prize money into the FA Cup revenue streams is quite new (2001/2002) and was a response to perceived declining interest by fans in the competition (Szymanski, 2001). A Cup draw matching a lower division team against a Premier League team also has a greater chance of being televised than a fixture between two lower division teams and this offers additional revenue-generating opportunities.

3 Model and Data

Szymanski and Smith (1997), Szymanski and Kuypers (1999) and Hall, Szymanski, and Zimbalist (2002) have outlined a simple model that relates team financial resources to team performance. In a panel setting for team i in season t:

$$\text{Revenue}_{it} = f(\text{team performance}) \quad (1)$$

$$\text{Team performance}_{it} = g(\text{relative wage bill}_{it}) \quad (2)$$

In this model, *relative wage bill* is the total team payroll for club i divided by the divisional average for season t. The wage bill is seen as a suitable proxy for team talent. Team performance can be measured as end-of-season league position. In his various papers, Szymanski uses log odds of position defined as $-\ln \frac{n}{93-n}$ for the English league as a whole comprising 92 teams across 4 divisions.

In a competitive market, players are paid according to their marginal revenues products and the total payroll of the team squad is simply the sum of wages which is then the sum of marginal revenue products. It is assumed that any externalities in player talents are fully dissipated into player salaries. We should note several problems with the assumptions that lie behind the measure of wage bills as a proxy for team talent resources. First, there is a concern with the available financial data that the payroll figures shown are actually total wages for the whole club, not just the players, and so include commercial, stadium and coaching staff. A further assumption needs to be made that payrolls on non-playing staff are a fixed proportion of payrolls of playing staff, but this cannot be verified. Second, transactions costs and player preferences for location and for matching with particular managers and coaches serve to inhibit arbitrage in the player labor market so differences in pay do not serve to encourage mobility so as to equalize marginal revenue products. Third, the labor market for players in the English leagues was subject to some restrictions in the 1970s, with full freedom of contract only enforced in 1978. Up to that point, clubs could keep a player's registration even if their contracts with players had expired, under the "retain and transfer" system. The principle of freedom of movement across football leagues within the European Union was only enforceable after the Bosman ruling of 1995. Despite these objections, Simmons and Forrest (2004) found that *relative wage bill* was significantly and positively related to team performance in the top divisions of England, Germany and Italy. In the case of Germany and Italy, the recorded wage bills were actual player payroll figures. However, Simmons and Forrest did not investigate the problem of causality between payroll and performance. Hall et al. (2002) performed Granger causality tests for the payroll-performance relationship in English football and found that payroll did indeed Granger-cause team performance.

An alternative model of football team performance was set out by Buraimo, Forrest and Simmons (2007):

$$\text{League position} = f(\text{relative wage bill}) \quad (3)$$

$$\text{Relative wage bill} = g(\text{relative revenue}) \quad (4)$$

$$\text{Relative revenue} = h(\text{market size}) \quad (5)$$

These equations can be combined to give a quasi-reduced form:

$$\text{League position} = f(\text{relative revenue, market size}) \quad (6)$$

In this variation of Szymanski's model, team performance again depends on relative wage bill of teams but this is now endogenous to relative revenues. Team revenues are quickly, and almost fully, transformed into spending on player talent.Indeed, in English football, the correlation coefficient between team revenues and wage bills is observed to be almost one (Buraimo, Simmons, & Szymanski, 2006). Relative revenues, which are team revenues divided by league or division average in a given season, are determined by market size.

Without a natural experiment, we cannot distinguish between the Szymanski model and the Buraimo model shown above. The payroll-performance equation is not identified. There are actually three problems that lead to lack of identification of the payroll-performance relationship First, the team performance equation may suffer from omitted variable bias in that some other component of team financial or playing resources is not captured by the model. Second, the relative wage bill or relative revenue variables my be subject to measurement error. Finally, team performance and team payrolls or revenues may be simultaneously determined. We shall focus here on the first of these problems and exploit the FA Cup as an additional source of team revenues and playing resources that can help teams augment their playing talent and improve their team strengths and consequent performances. Industry folklore suggests an externality between FA Cup success and League success, as winning Cup games builds team momentum and crowd support that spills over into League performances. If this momentum effect is valid, it should hold independently of team revenue generation so that:

$$\text{League position} = f(\text{relative revenue, market size, FA Cup progress}) \quad (7)$$

But performance spillovers is not the mechanism that we have in mind for FA Cup draws to impact on league positions. Rather, we propose to use various dimensions of FA Cup draws as instruments for team revenues. It is the random draw that creates possibilities for additional revenue generation, and these opportunities can occur independently of the actual outcomes of Cup games, such as a small team beating a big team in what is termed a "giant-killing" exploit. The random assignment of teams in Cup draws means that extra revenue earned from playing a big team in a given round is itself a source of random variation in team resources. Hence, we use FA Cup draw measures as instruments for endogenous team revenues in order to resolve a potential missing variable problem for our performance-revenue relationship. We can, of course, test for additional impacts of FA Cup success on League positions through momentum created by FA Cup performance but these effects occur for a given amount of revenue.

Hence our model will be

$$\text{League position} = f(\text{relative revenue, FA Cup success, team fixed effects}) \quad (8)$$

where league position is simply the end-of-season divisional ranking with 1 for first, 2 for second and so on;

$$\begin{aligned}\text{Relative revenue} = g(&\text{FA Cup draw, team fixed effects,}\\ &\text{lagged relative revenue})\end{aligned} \quad (9)$$

In this adaptation of the model of Buraimo et al. (2007) we subsume market size under team fixed effects. FA Cup draws will be captured by three variables. We take rounds 3–5 and create dummy variables for participation in each of these rounds. Round 3 is where teams from the top two divisions first enter the competition by right. It is very rare for teams from the bottom two divisions to progress beyond round 5. Round itself is clearly not exogenous since achieving any particular round depends on team performance. However, *Round* × *home* denotes round reached interacted with a home draw. Since *home* is determined by random drawn and there is a well-known home advantage that therefore affects the probability of winning that Cup tie, irrespective of opponent identity and status.

Round × *opponent size* captures the predicted gain in revenue from meeting a particular opponent. *Opponent size* is measured by stadium capacity and not League (or Cup) attendance. Attendances could well be endogenous to Cup progress and success while ground capacity is pre-determined before League and Cup competitions develop and can be taken as exogenous. We note the reconstruction of most English football stadia following the Taylor Report of 1989. Following the Hillsborough stadium disaster of April 1989 in which 96 fans lost their lives due to crowd congestion and crushing at an FA Cup semi-final at Sheffield Wednesday's ground, clubs were instructed by Government to refit their stadia to all-seater specifications. This led to smaller stadium capacity figures, but increased gate revenues, for most clubs. We made every effort to locate accurate stadium capacity figures especially during the refitting of stadia in the early 1990s, using *Rothmans Football Yearbook* and a thorough online search for club histories. Finally, we combine *round* reached with a dummy variable for opponent from the top division, *round* × *top team*, to capture the boost to revenues from meeting a big team. Typically, clubs in the top division bring higher revenues to Cup games than teams from lower divisions.

Hence, our full model is:

$$\text{League position} = f(\text{relative revenue, FA Cup success, team fixed effects}) \tag{10}$$

$$\text{Relative revenue} = g(\text{Round} \times \text{home, Round} \times \text{opponent size, Round} \times \text{top team, lagged relative revenue, team fixed effects}) \tag{11}$$

The use of round interaction terms means that the impact of an instrument is switched on according to the round reached. This model is estimated by instrumental variables with team fixed effects at each stage. The FA Cup draw variables serve as instruments and we test for under-identification, weakness of instruments, and for over-identification. We estimate the model separately by division and then as a pooled regression combining divisions. We then contrast an OLS league position fixed effects model with the IV version to check that the impacts of revenue on team performance are different when we allow for FA Cup draw effects on revenues. If the FA Cup draw is irrelevant to team revenues and team performances then the revenue effects on performance should be similar in both OLS and IV models. If the FA Cup draw adds to relative revenues for those teams that encounter big teams then we predict that this boost to revenues will enhance team performances to a greater extent than if FA Cup revenues are excluded.

Our data on team revenues come from Szymanski and Kuypers (1999), Stefan Szymanski and various issues of the Deloitte Annual Review of Football Finance. These data are taken from company accounts. Our sample begins at 1971/1972 and ends in 2007/2008 for the top two divisions but the best coverage for all divisions in from 1991/1992 to 2000/2001 with incomplete reporting for all divisions before 1991 and no records at all for the bottom two divisions after 2001. The fourth tier of the League is particular patchy in coverage with just three to six teams per season offering revenue data in the 1970s. Empirical results from the fourth tier must then be treated with caution. However, t-tests of mean finish positions for tiers 3 and 4, comparing teams with reported revenues and those without fail to reject the null hypothesis of equality of positions. This gives us some encouragement to persist with the relative revenue measure even where coverage is thin.

Details of FA Cup performance, including identity of opponents, round played and results, were extracted from www.soccerbase.com. Stadium capacity figures were obtained from editions of *Rothmans Football Yearbook* and cross-checked against online records of club histories, with particular attention given to the stadium reconstruction period of the early 1990s.

4 Empirical Results

Table 2 shows mean revenues, both as totals and relative to divisional averages, for the four divisions according to progress through the FA Cup. We show three categories of revenue: by highest round of FA Cup achieved, whether or not teams play top division opponents in a particular round and whether or not teams defeat top division opponents in a particular round.

In broad terms, it appears that teams that progress to round 5 or beat top division teams generate higher relative revenues than teams who exit the FA Cup at round 3 or earlier. For tiers 1 and 2, either playing or beating a top division opponent in round 5 is associated with greater total and relative revenue associated with being eliminated at the earliest opportunity in round 3. In the bottom two divisions, either playing or beating a top division opponent in round 4 is associated with greater total and relative revenue associated with being eliminated before round 3. It is also notable that teams from the bottom two divisions that actually reach round 3 have higher revenues, on average, than teams that exit prior to this stage. However, the reported revenues in each FA Cup category do not increase monotonically by round.

Table 2 Team revenues by division and FA Cup round, £m

	Round 1 or 2	Round 3	Round 4	Round 5
Tier 1				
Round reached		19.80	23.10	34.20
		(0.835)	(0.901)	(1.168)
Play top division team		16.50	26.20	34.20
		(0.850)	(0.949)	(1.230)
Beat top division team		30.50	46.10	37.40
		(1.238)	(1.217)	(1.359)
Tier 2				
Round reached		4.96	6.78	6.33
		(0.921)	(1.064)	(1.109)
Play top division team		5.41	6.36	6.19
		(0.931)	(1.106)	(1.081)
Beat top division team		7.10	7.03	8.07
		(1.160)	(1.238)	(1.308)
Tier 3				
Round reached	1.43	1.86	1.44	1.91
	(0.937)	(1.068)	(0.999)	(1.206)
Play top division team		2.20	1.37	2.10
		(1.134)	(0.998)	(1.309)
Beat top division team		1.74	2.09	
		(1.036)	(1.166)	
Tier 4				
Round reached	0.82	1.15	1.24	
		(0.930)	(1.096)	(1.385)
Play top division team		1.44	1.23	
		(1.119)	(1.490)	

Note: relative revenues, scaled by divisional averages for a given season, are shown in parentheses. *Round reached* denotes the highest round achieved i.e. no further progress beyond the specified round.

Table 3 Impacts of relative revenues on League position

	Tier 1	Tier 2	Tier 3	Tier 4
OLS fixed effects only	−3.136 (2.70)	−5.177 (6.42)	−5.416 (5.24)	−7.978 (3.21)
OLS fixed effects plus Cup win effects	−2.995 (2.78)	−5.269 (6.71)	−5.067 (5.09)	
Beat top team × round 3	−1.033 (1.80)	−0.303 (0.32)	−1.420 (1.17)	
Beat top team × round 4	0.023 (0.04)	0.337 (0.33)	−4.398 (5.84)	
Beat top team × round 5	−1.207 (3.08)	1.839 (1.34)	−0.517 (0.17)	

Note: Absolute t statistics in parentheses.

We begin our regression analysis in Table 3 with a fixed effects OLS model of team finishing positions, estimated separately by division, with relative revenues as the sole exogenous regressor. The impact of relative revenue on League position is highest in the bottom division and greatest in the top division. Hence, a given increment in relative revenue is accompanied by greater returns in terms of League performance in the bottom tier. This appears to be plausible as the bottom tier contains a set of relatively small, somewhat homogeneous clubs, all with limited access to revenues. In this lowest division, a modest increase in relative revenues can deliver a substantial gain in League position. In the top tier, revenues are more highly dispersed than at the bottom. The standard deviation of relative revenue in the top tier is 0.65 compared to 0.42 for the lowest division. It is also apparent that the dispersion of relative revenues in the top division, has increased over time from a standard deviation of 0.48 before 1992, at which point it became restructured as the FA Premier League, to 0.76 under the new branding. This supports the finding of increasing dispersion of revenues stated by Dobson & Goddard (2004). In the post-1992 Premier League a given increase in relative revenue is associated with a smaller impact on League position.

Next, we add FA Cup momemtum effects to the OLS League position model. Momentum is captured by a set of dummy variables, *FA Cup progress*, defined as round reached interacted with victory over a top division team: *Beat top team × round 3* through to round 5. The bottom division is excluded from this model since the number of cases where a team from this division defeated a team from the top division was two for round 3 and just one for round 4. For the top division, we see that defeating a same-division rival in round 5 is associated with a modest bur significant improvement in League position of 1.2 places, controlling for unobserved team fixed effects. In tier 2, we see no significant effect at all from defeating top division teams on League position. In tier 3, we find a substantial, significant effect on League position from beating a top division team in round 4, with a 4.4 place improvement.

However, the performance-revenue relationships shown in the OLS estimates seen in Table 3 are not identified. Table 4 shows estimates of a two stage least squares model, pooled across divisions, in which FA Cup draw covariates are instruments. It is clear that when these instruments are included alongside relative revenue in an augmented OLS fixed effects model then not one of these instruments has a significant coefficient. Hence, a favorable draw against a top tier opponent, cannot

Table 4 Pooled estimates of a 2SLS model of league positions by division

	Tier 1 1971/1972 to 2007/2008	Tier 2 1971/1972 to 2007/2008	Tier 3 1971/1972 to 2000/2001	Tier 4 1971/1972 to 2000/2001
Relative revenue 1st stage				
Round 3 home	–0.008 (0.25)	–0.053 (1.53)	–0.025 (0.47)	**0.143 (2.20)**
Round 4 home	0.073 (1.84)	–0.009 (0.20)	–0.033 (0.45)	0.162 (1.42)
Round 5 home	0.150 (1.84)	–0.053 (1.12)	0.218 (1.48)	
Round 3 opponent capacity	**–0.003 (2.03)**	0.001 (0.56)	–0.001 (0.75)	–0.002 (0.34)
Round 4 opponent capacity	0.001 (0.58)	0.002 (1.09)	**0.001 (1.96)**	0.003 (1.56)
Round 5 opponent capacity	–0.002 (1.03)	0.002 (0.67)	**–0.011 (2.29)**	
Round 3 top opponent	**0.108 (1.96)**	–0.020 (0.40)	**0.204 (2.01)**	0.090 (0.81)
Round 4 top opponent	–0.029 (0.50)	0.039 (0.57)	–0.140 (1.22)	0.404 (1.84)
Round 5 top opponent	0.057 (1.13)	–0.034 (0.34)	**0.627 (3.15)**	
Lagged relative revenue	0.332 (2.68)	0.041 (1.40)	0.056 (2.41)	0.062 (2.26)
League position 2nd stage				
Relative revenue (IV)	–4.889 (5.37)	–5.178 (8.74)	–6.072 (10.58)	–8.354 (13.26)
Beat top team × round 3	–0.976 (1.74)	–0.502 (0.56)	**–3.169 (2.45)**	3.078 (1.25)
Beat top team × round 4	0.348 (0.57)	0.600 (0.61)	**–4.393 (2.09)**	
Beat top team × round 5	–0.832 (1.39)	0.637 (0.48)		

Notes: $N = 1739$, Kleibergen-Paap LM test for underidentification = 127.7 ($p = 0.000$), Kleibergen-Paap Wald F test for weak identification = 2.86, Hansen J test for overidentification = 47.87 ($p = 0.09$).

be taken, in itself, to have a "momentum" effect on League performance by raising confidence and "form" of teams. The FA Cup can have an impact on League performance through revenue-generation and the dissipation of revenues into team strength. This can be achieved by retention of players who might otherwise leave or by hiring new players to reinforce team resources.

Table 4 shows estimates of our two-stage least squares model, pooled across divisions. The results do not give much support for the application of FA Cup variables as instruments. Few of the instruments have significant coefficients. In the case of tier 2, none of the instruments returns a significant coefficient.

Nevertheless, there are some signs that FA Cup performance in some divisions does have an impact on League success. In the top tier, we see that meeting a rival top division team in round 3 adds significantly to relative revenue. However, this is offset by the negative effect of capacity of the opponent. For top division capacity levels of around 30,000, which corresponds to an average ground size over the sample period for this level, the gain to revenue from meeting a top team is wiped out by the fact that a sizeable opponent is confronted. This suggests that our capacity instrument actually conflates size of opponent with probability of winning. A big team with higher ground capacity is harder to beat. Moreover, if this big team comes from one's own division and wins the Cup tie then that rival advances to later rounds to accumulate more revenue from gates, broadcast viewing and prize money. Meanwhile, the defeated team is forced to "concentrate on the League" to use the industry cliché but with fewer relative resources.

For the third division, it appears that playing a top division team in round 5 adds significantly to relative revenue. This impact is reduced as capacity of top division opponent is increased, but never enough in the sample to eliminate the positive effect entirely. A tier 3—tier 1 match in round 5 is not a frequent event and only occurs 18 times over a 30 year period in our sample i.e. once every two seasons. But when it does happen a tier three team benefits from extra revenue streams. Of course, top division teams are on average more likely to progress to round 5 than lower division teams. Hence, conditional on reaching round 5, the probability of being drawn against a top team is quite high (0.64 in our sample).

The diagnostic test for underidentification rejects the null while the Hansen J overidentification test does not reject the null that the model is identified. However, the Kleibergen-Paap LM test reveals a problem of weak identification. This is not surprising as several of the instrumental variables have insignificant coefficients.

The impacts of relative revenue on team League performance are shown in Table 4 for the unrestricted case of relative revenue coefficient inequality. The coefficients are higher than in the OLS fixed effects case but are closer together in magnitude. F-tests suggest coefficient equality for tiers 1–3, but not 4, and so we proceed to restrict the model to impose equal impacts of relative revenue on performance across the top 3 divisions. The constrained coefficient on relative revenue in the league position equation is –6.339 with a t-statistic of 11.96. With the coefficient equality restriction imposed, the null hypothesis of weak identification is rejected, according to the Kleibergen-Paap critical values for their test, applied to a test statistic of 19. Also, the test for underidentification again rejects the null while the Hansen J test fails to reject the null of identification at 5 % (p value of 0.06).

5 Discussion and Conclusions

Overall, our results show a larger impact of relative revenue on team league performance when revenues are endogenous to FA Cup performance. At first glance, this is not consistent with omitted variable bias apparent in estimates of wage equations, for example, where number of years of education is an imperfect proxy for worker ability. In that case, the impact of the endogenous regressor on the dependent variable is biased upwards in OLS, and instrumentation lowers the coefficient. Here, the coefficient rises once we introduce our set of FA Cup instruments. The apparent paradox can be resolved by thinking of the performance-revenue relationship as a long-run equilibrium relationship that is subject to shocks from unforeseen FA Cup performance. When these shocks are accounted for, League performance is raised above the normal level for a given set of inputs. The failure of our most general model to pass the weak identification test does, however, suggest caution in making inferences. Nevertheless, all instrumental variable specification tests pass when the lowest division is excluded.

We found some evidence of a momentum effect of FA Cup success on team league positions, given revenue, especially for tier 3. Impacts of FA Cup draws on revenue are entirely absent for tier 2. For teams at this level, the FA Cup appears orthogonal to their objectives which appear either to gain promoted to the top tier or to retain divisional status. Neither a favorable FA Cup draw nor a successful Cup run help tier 2 teams to make progress in the League. For tier 3 teams, we do observe some impacts of FA Cup draws and momentum on League success. In particular, meeting a moderately sized top division team, with 25,000 ground capacity, in round 5 generates a positive impact on League position of 2.3 places. Beating a top division opponent in rounds 3 or 4 is predicted to generate an improvement in League position for a tier 3 team of 3.2 and 4.4 places, *cet. par.*

The limited success of our FA Cup draw instruments in explaining variation in team revenues suggests that revenue-sharing in the FA Cup does not have a great deal of impact on League performance. For tier 2 teams, several of which are aspiring to a prize of a large revenue bonus (estimated by industry experts at around £60m), the FA Cup is simply irrelevant. If revenue-sharing were to be abandoned in the FA Cup then teams in the bottom two divisions would lose some of the opportunity to enhance their revenues and improve their League performance. At present, the FA Cup combines elements of prize money and revenue-sharing and both of these serve as important incentives for teams in the third tier of English football. In the bottom fourth tier, the prospect of a home draw in the third round, after many divisional rivals have been eliminated, gives a revenue advantage over competitors from the same division.

We lack revenue data for teams in the bottom two divisions after 2001. Since then, English football has adopted the "transfer window" practice by which teams can only trade players with permanent contracts within a season during the month of January. Before 2001, the deadline for player transfers in a given season was at the end of March. The current transfer window serves to restrict trades and, in particular, means that teams that make progress in the FA Cup cannot convert the extra revenues into player acquisitions financed by transfer fees selling clubs. We predict, therefore, that any beneficial effects of FA Cup draws on revenues and hence League performance would be reduced as this restriction on player trades comes into place.

References

Borland, J., & Macdonald, R. (2003). The demand for sports. *Oxford Review of Economic Policy, 19*, 478–502.

Buraimo, B., Forrest, D., & Simmons, R. (2007). Freedom of entry, market size and competitive outcome: Evidence from English soccer. *Southern Economic Journal, 74*, 204–213.

Buraimo, B., & Simmons, R. (2009). A tale of two audiences: Spectators, television viewers and outcome uncertainty in Spanish football. *Journal of Economics and Business, 61*, 326–338.

Buraimo, B., Simmons, R., & Szymanski, S. (2006). English football. *Journal of Sports Economics, 7*, 29–46.

Dobson, S., & Goddard, J. (2001). *The economics of football*. Cambridge: Cambridge University Press.

Dobson, S., & Goddard, J. (2004). Revenue divergence and competitive balance in a divisional sports league. *Scottish Journal of Political Economy, 51*, 359–376.

Fort, R., & Quirk, J. (1995). Cross-subsidization, incentives, and outcomes in professional team sports leagues. *Journal of Economic Literature, 33*, 1265–1299.

Hall, S., Szymanski, S., & Zimbalist, A. (2002). Testing causality between team performance and payroll: The cases of major league baseball and English soccer. *Journal of Sports Economics, 3*, 149–168.

Késenne, S. (2000). Revenue sharing and competitive balance in professional team sports. *Journal of Sports Economics, 1*, 56–65.

Késenne, S. (2007). *The economic theory of professional team sports: An analytical treatment.* Cheltenham: Edward Elgar.

Maxcy, J. (2009). Progressive revenue sharing in major league baseball: The effect on player transfers and talent distribution. *Review of Industrial Organization, 35*, 275–297.

Noll, R. (2002). The economics of promotion and relegation of in sports leagues: The case of English football. *Journal of Sports Economics, 3*, 169–203.

Rosen, S., & Sanderson, A. (2001). Labour markets in professional sports. *Economic Journal, 111*, F47–F68.

Rottenberg, S. (1956). The baseball players' labor market. *Journal of Political Economy, 64*, 242–258.

Simmons, R., & Forrest, D. (2004). Buying success: Relationships between team performance and wage bills in the U.S. and European sports leagues. In R. Fort & J. Fizel (Eds.), *International sports economic comparisons* (pp. 123–140). Westport, CT: Praeger.

Szymanski, S. (2001). Income inequality, competitive imbalance and the attractiveness of team sports: Some evidence and a natural experiment from English soccer. *Economic Journal, 111*, F59–F84.

Szymanski, S. (2003). The economic design of sporting contests. *Journal of Economic Literature, 51*, 1137–1187.

Szymanski, S., & Késenne, S. (2004). Competitive balance and gate revenue-sharing in team sports. *Journal of Industrial Economics, 51*, 513–525.

Szymanski, S., & Kuypers, T. (1999). *Winners and losers: The business strategy of football.* London: Viking Press.

Szymanski, S., & Smith, R. (1997). The english football industry: Profiot, performance and industrial structure. *International Review of Applied Economics, 11*, 135–153.

Commercial Football and the Economic Cycle

Stefan Szymanski

Abstract This chapter examines the financial structure of football clubs and considers how this relates to the economic cycle. Some evidence is provided on trends in attendance at European football leagues. Regression evidence suggests that there is no statistically significant link between economic growth and attendance growth, but that there does appear to be an "international tournament cycle", whereby attendance seems to rise in the year preceding either the FIFA World cup or the UEFA European Championships.

1 Introduction

We are accustomed to taking about financial crises in the football industry, but in the last year we have witnessed a financial crisis on a much larger scale. The bankruptcy of Lehman Brothers in September 2008 signalled the meltdown of the global banking industry. Banks were forced to write down huge losses mostly due to lending in the subprime mortgage market and in the effort to rebuild their balance sheets reduced lending to commercial sector. The loss of confidence, not least among consumers, caused a sharp decline in retail sales and a build up of inventories which in turn led to reductions in output and layoffs across Europe and the US.

On the face of it one might expect falling consumer demand to have a significant effect on the demand for football, or for commercial sport in general. Spending on football tickets, merchandising and TV subscriptions is a form of consumer spending, and therefore the revenue generating potential of football clubs is likely to come under some pressure. It was therefore a cause of some surprise when Real Madrid acquired Ronaldo and Kaka in the summer of 2009, spending over €200 million at a time when most other businesses were seeking to keep costs down. During the same period Manchester City, owned by investors from Abu Dhabi, spent around €120 million (net) on acquiring players.

S. Szymanski (✉)
Cass Business School, City University London, London, EC1Y 8TZ UK
e-mail: szy@city.ac.uk

S. Butenko et al. (eds.), *Optimal Strategies in Sports Economics and Management*,
DOI 10.1007/978-3-642-13205-6_10,

The paper is set out as follows. In the next section the economics of football management is discussed, then in Sect. 3 the relationship between attendance demand and the current economic crisis is considered while Sect. 4 looks at the relationship between league attendance demand and GDP for a number of European countries over the last two decades. Section 5 concludes.

2 The Economics of Football Management

The business model of football is relatively simple to understand. Clubs invest in players and facilities in order to attract paying fans. Fans are sensitive to team success and therefore better teams tend to generate more income. To generate more success teams need to hire the best players; since player skills are readily observable and there exists a large number of buyers and sellers, increases in wage spending tend on average to lead to proportionate increases in performance.

These two propositions can be illustrated using financial data for the Spanish leagues over the period 1995–2004.[1] Figure 1 shows the relationship between wage spending and league position over time. The absolute level of wage spending has been driven by the very high rate of salary inflation in football which averaged in the region of 20% per year during this period in Spain (this is not been untypical of salary inflation across the major European leagues for many years now). But what matters in determining league performance is relative spending-teams that spend above average compared to their rivals tend to outperform. Thus R^2 of the regression in Fig. 1 is 0.581, meaning that approximately 58% of the variation in

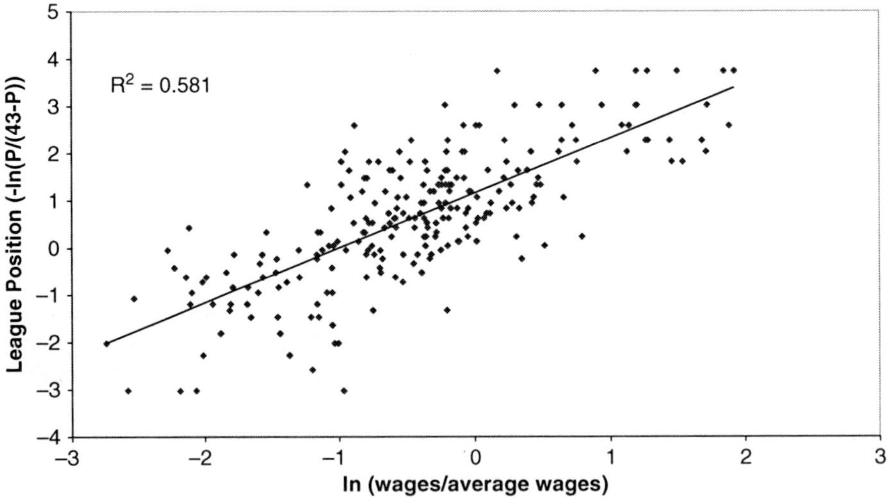

Fig. 1 Wage spending and league position, Spain, 1995–2004

[1] The data is that used in Garcia-del-Barrio and Szymanski (2009).

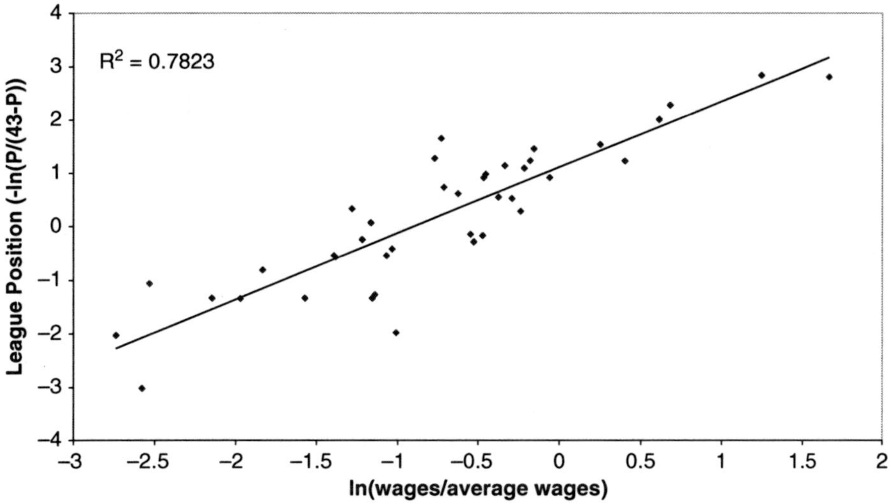

Fig. 2 Wage spending and league position, Spain, 1995–2004

league position can be explained by wage expenditure. Other factors matter, not least of which is luck, which take the form of player injuries, poor refereeing or other random events, but this kind of randomness tends to even itself out over time. This point is illustrated in Fig. 2 which averages the performance of individual teams over time, and shows that relative wages are even more powerful in explaining performance over time, the R^2 indicating that almost 80% of the variation in league position can be explained by relative wage expenditure.

Figures 3 and 4 illustrate the same points for the relationship between performance in the league and revenues. In any given season league position can account for about 70% of the variation in operating incomes of the clubs, but over the longer term it explains almost 90%. Given the alleged loyalty of fans to their team, it is perhaps surprising to see so much of the variation in revenues explained only by success. Some fans might desert to a more successful team if their club is unsuccessful, but it seems more likely that fans might just stop going to watch games (and/or buying merchandise such as shirts). And even if existing fans may be committed, each season new fans enter (often as children who will follow their first team for the rest of their lives) and these new fans tend to be attracted by teams that are currently successful.

While these figures illustrate the case of Spanish football, much the same results can be found for other football leagues and leagues in other sports (see for example Szymanski and Smith (1997) Forrest and Simmons (2002) and Szymanski (2003)). Hall, Szymanski, and Zimbalist (2002) also tested for the direction of causality (in case it might be thought that wages are a reward for wins rather than a measure of playing talent) for the case of English football and found that wages did indeed appear to cause higher league positions.

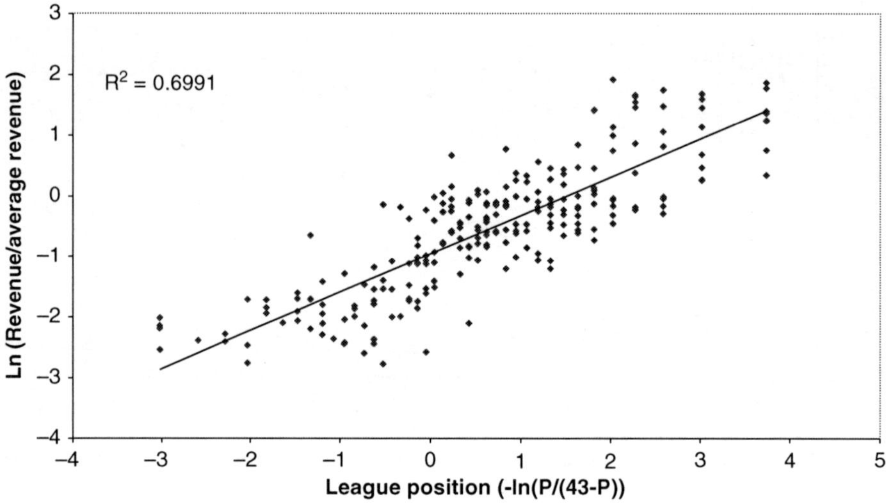

Fig. 3 League position and club operating income, Spain, 1995–2004

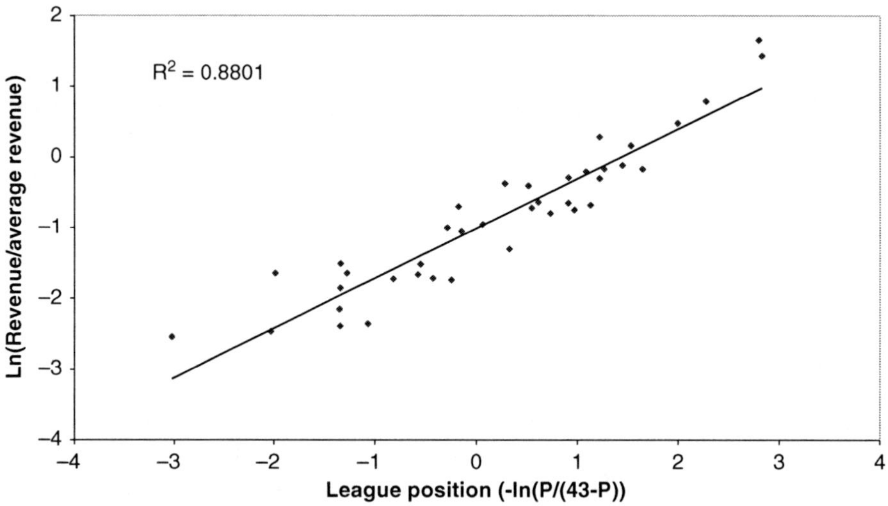

Fig. 4 League position and club operating income, Spain, 1995–2004

Given there exists a relatively stable relationship between revenues and wages, through player spending, it might be argued that clubs can effectively choose a level of profitability contingent on the amount of talent they choose to hire. However, Garcia-del-Barrio and Szymanski (2009) show that the league positions achieved by clubs in Spanish football are consistent with win–maximizing behavior—meaning that teams spend all available income on players in order to achieve the highest league position possible, rather than maximizing profits. Even in the English league, where clubs sustain a significantly higher level of operating profits and several clubs have raised capital on the stock exchange from investors who are presumably driven by profit, they still found that behavior seemed closer to win maximizing conduct than profit maximization. In part this may be attributable to competition to win places in European competition which can generate substantial additional income, and the fear of relegation which is likely to threaten any club that limits its player spending.

This is in stark contrast to the American sports model, where clubs operate in closed leagues (no promotion and relegation), with considerable restraints of player spending (e.g. roster limits and salary caps), and substantial revenue redistribution (e.g. equal sharing of broadcast income, gate revenue sharing). Team owners there are generally thought to be profit maximizers and indeed to earn substantial profits from the operation of sports teams.

If we consider this model a little more formally, we can relate position to revenues and wages through a pair of equations:

$$P_i = a + b\frac{w_i}{\overline{w}} \tag{1}$$

$$\frac{R_i}{\overline{R}} = c + dP_i \tag{2}$$

Where P is league position, R is revenue, w is wage spending, and a bar represents the league average. If we substitute out for P_i, then we obtain a relationship between revenue and wages. If profits are simply $R - w$, and clubs spend up to the points where profits equal zero (win maximization), we obtain the following equation for wages and revenues:

$$w_i = \frac{\overline{w}\overline{R}(ad + c)}{\overline{w} - bd\overline{R}} \tag{3}$$

$$R_i = \frac{\left[\overline{w} - bd\overline{R} + bd\right]\overline{R}(ad + c)}{\overline{w} - bd\overline{R}} \tag{4}$$

The point of this simple model is to illustrate that revenues will depend significantly on the conduct of other teams in the league and on the underlying parameters, a, b, c and d. Clubs make their financial plans during the summer preceding the football season. A financial plan will involve a level of player spending intended to achieve a given league position and a level of revenue generation consistent with these intentions. The plan would require the team to form expectations about the level of wage spending of the other teams—if rivals turn out underspend significantly compared to expectations the implication is that revenue and therefore profit will be higher than expected (because performance in the league will be better than expected). Similarly, if rivals overspend relative to expectations then revenues, profits and league performance will be lower—indeed the club will makes losses in this case. We may call this the "wage risk". This kind of risk is to some extent limited by the fact that managers have a fairly good idea about the plans of rival teams and most spending decisions are reasonably transparent. Although it is possible to be surprised by a surge in transfer activity as the transfer deadline approaches, it should not in fact prove too difficult to control this kind of risk.

On the other hand the "revenue risk" lies in the possibility club misjudges the parameters of the revenue Eq. (2).[2] Teams may rely on fixed revenues which are independent of performance (these might include long term sponsorship contracts and rental income from the use of facilities) and a number of revenues that are sensitive to performance (ticket sales, merchandising sales and revenues from competitions such as the Champions League and domestic cup competitions). Revenue risks are thus in part dependent on prevailing macroeconomic conditions which influence spending patterns that may be directly or indirectly related to the performance of the team. In extreme cases revenue risks can arise from the bankruptcy of customers who are no longer able to honor their contracts (as has happened on a number of occasions in relation to broadcasting contracts).

3 The Impact of the Current Financial Crisis

In this paper I will consider attendance related revenue risk. To do so I will focus mainly on the attendance database at http://www.european-football-statistics.co.uk. This provides club by club attendance data for most European leagues over a range of years, although I will use the average attendance for the top division in each league. The impact of promotion and relegation is inevitably problematic in any

Table 1 Percentage change over previous year, total attendance in top division and GDP 2009, western Europe. Source: http://www.european-football-statistics.co.uk and International Monetary Fund, World Economic Outlook Database, October 2009

Country	Percentage of change in total attendance	Percentage of change in GDP
Sweden	–42.37	–4.83
Switzerland	–36.46	–1.95
Iceland	–30.66	–8.51
Finland	–21.87	–6.36
Portugal	–13.58	–3.00
Norway	–12.42	–1.91
France	–9.83	–2.36
Spain	–7.04	–3.77
Belgium	–5.81	–3.18
UK (England)	–3.51	–4.38
UK (Scotland)	–0.76	–4.38
Denmark	–0.56	–2.43
Netherlands	9.37	–4.17
UK (Wales)	9.76	–4.38
Germany	10.49	–5.30
Italy	12.67	–5.14
Greece	37.17	–0.75

[2] Note that this is an example only, and in reality the revenue function is more complex, so that estimating the true parameters is likely to be even more difficult than the model suggests.

analysis of this kind. Relegation leads to particularly large changes in attendance, and a balanced panel of teams would necessarily involve quite large changes in the relative quality of the teams in the sample from year to year. On the other hand, if, as here, one focuses on the level of the league, then a promoted team may bring a quite different level of support (higher or lower) in to the league, causing sharp variations in average attendance from year to year.

Table 1 illustrates the sensitivity of attendance to the economic cycle in western European leagues. All countries experienced a sharp decline in GDP in 2009 and 12 out of 17 experienced a fall in attendance in the top division.

The picture is less clear when we examine the countries of eastern and central Europe. Although the economic collapse was even greater than in most western European countries (with the exception of Poland), only five of the eleven saw a decline in attendance.

4 Trends in Europe Across Time

GDP is clearly not the only factor that influences demand for attendance at football stadiums. In general we would expect prices to play an important role, as well as team performance, and more generally the drawing power of the team (e.g. population in the local area, the intensity of attachment to the team and so on).

There are a large number of demand studies in team sports (see e.g. the reviews by Borland and Macdonald (2003) and Simmons (2006)). Most of these studies use cross-sections or have a short time series dimension. US studies of baseball involving longer time series have tended to focus on issues such as the uncertainty of outcome hypothesis (see e.g. Schmidt and Berri (2001) and Humphreys (2002) and include few other explanatory variables. Studies of demand for English football over lengthy time periods (e.g. Dobson and Goddard (1995) and Simmons (1996)) have included various economic and demographic variables such as ticket prices and local population, employment and occupation, wage levels and unemployment. These studies, like many others, find that demand tends to be price inelastic, while income related variables are seldom significant. Estimates by Simmons suggest that demand is in general inversely related to price, while he observes that unemployment effects are difficult to capture because of their collinearity with wages. Dobson and Goddard (1996) use unemployment as a proxy for incomes and find a long run relationship with attendance while prices exert a short term influence on demand but there is not long term relationship.

In this paper the focus is on demand for attendance across a number of European countries, and there are no time series of ticket prices available. The objective is to identify whether there is any relationship between attendance and variations in GDP over the economic cycle. Table 2 shows the changes in attendance and GDP per capita for western European nations over the sample period, which ends in 2009, meaning the season 2008/2009 (except for Cyprus), and starts either in 1993 or 2001 in a majority of cases. GDP per capita is preferred to GDP alone since it seems

Table 2 Percentage change over previous year, total attendance in top division and GDP 2009, central and eastern Europe. Source: http://www.european-football-statistics.co.uk and International Monetary Fund, World Economic Outlook Database, October 2009

Country	Percentage of change in total attendance	Percentage of change in GDP
Latvia	–38.00	–18.00
Czech Republic	–14.72	–4.32
Bulgaria	–8.07	–6.50
Russia	–3.06	–7.55
Poland	–0.68	0.97
Croatia	10.11	–5.24
Lithuania	16.39	–18.50
Slovak Republic	16.73	–4.66
Hungary	25.35	–6.73
Romania	25.46	–8.46
Slovenia	40.44	–4.72

more likely to capture the income effects that might influence demand for football attendance. The GDP data is for the calendar year (for 2009 the data represents IMF estimates). In general this means that the GDP figures coincide with only half of the football season in question (i.e. if the season runs from August/September to May). A number of alternative versions using a weighted average of GDP in consecutive years were considered, but none fitted the data any better. To the extent that demand is related to income not yet received, it may be that decisions on attending football matches are more dependent on prospective income rather past income. It would be interesting to develop the approach considered here by examining quarterly data where that is available. Indeed it might be conjectured that seasonal patterns of income have an important influence on demand for football since economic activity tends to be greater during the football season.

Table 3 ranks countries by their change in total attendance. The correlation coefficient between the first two columns is 0.42, suggesting that there is some relationship in cross section between growth in demand and growth in GDP per capita. The next column reports the correlation between attendance growth and GDP per capita for each country in the sample. Here the figures show no overall pattern—with seven out of seventeen cases showing a negative correlation over time between GDP per capita and attendance.

Table 4 shows the changes in attendance and GDP per capita for central and eastern European nations over the sample period. If anything the picture is even more random than the case of western Europe. Not only are individual countries as likely to exhibit negative correlation between GDP per capita and attendance, but the correlation coefficient between the first two columns is only 0.09.

These descriptive statistics do not suggest a very strong cyclical relationship between GDP and football attendance in Europe, but to test the matter more fully a regression approach was adopted. The data is an unbalanced panel and therefore models were tested with both country and period fixed effects were included,

Table 3 Percentage change over sample period, total attendance in top division and GDP per capita, western Europe. Source: http://www.european-football-statistics.co.uk and International Monetary Fund, World Economic Outlook Database, October 2009

Country	Percentage change in total attendance (%)	Percentage change in GDP per capita (%)	Correlation of change in attendance and change in GDP	Min year	Max year
Italy	−43.06	11.53	−0.40	1993	2009
UK Scotland	−19.01	21.37	0.60	1997	2009
France	−15.89	3.82	0.57	2001	2009
Cyprus	−3.16	17.02	−0.03	2001	2008
UK Wales	22.45	25.06	−0.15	1996	2009
Belgium	27.43	6.77	0.22	2001	2009
Spain	32.36	4.44	0.14	2001	2009
Iceland	50.52	8.77	0.27	2001	2009
Finland	75.78	52.94	−0.10	1993	2009
Austria	82.69	9.64	0.42	2001	2009
Portugal	99.49	−1.22	0.30	2001	2009
Switzerland	122.63	15.90	0.30	1999	2009
Germany	150.10	19.11	−0.22	1993	2009
Greece	153.44	57.88	−0.03	1993	2009
UK England	181.02	37.20	0.32	1993	2009
Norway	194.76	8.62	0.36	2001	2009
Denmark	208.37	26.84	−0.08	1993	2009
Sweden	269.94	39.00	0.44	1993	2009
Netherlands	409.84	31.17	0.17	1993	2009

Table 4 Percentage change over sample period, total attendance in top division and GDP per capita, central and eastern Europe. Source: http://www.european-football-statistics.co.uk and International Monetary Fund, World Economic Outlook Database, October 2009

Country	Percentage change in total attendance (%)	Percentage change in GDP per capita (%)	Correlation of change in attendance and change in GDP	Min year	Max year
Bulgaria	−73.58	45.02	−0.14	2001	2009
Hungary	−72.11	57.98	−0.22	1994	2009
Latvia	−22.70	38.51	0.34	2001	2009
Czech Republic	−22.41	56.54	0.04	1993	2009
Slovak Republic	−7.75	80.99	−0.17	1995	2009
Slovenia	−5.86	28.40	−0.35	2001	2009
Croatia	0.89	28.84	0.00	2001	2009
Romania	9.18	36.90	−0.36	2002	2009
Ukraine	80.20	14.84	−0.16	1993	2007
Poland	193.43	73.62	0.29	1996	2009
Russia	217.92	66.94	−0.39	1994	2009
Lithuania	430.52	40.25	0.03	2001	2009

Table 5 Regression of changes in annual attendance

	All countries	Excluding east	Excluding east and small
GDP per capita	–0.178	0.873	0.927
	0.500	0.864	0.964
No World or Euro	–0.106[a]	–0.072[b]	–0.079[b]
championship	0.041	0.044	0.048
World or Euro host	–0.022	–0.019	–0.015
	0.146	0.121	0.122
World or Euro winner	0.584[c]	0.579[c]	0.583[c]
	0.169	0.139	0.141
Observations	382	240	209
R^2	0.108	0.117	0.125
fixed effects	yes	yes	Yes
year effects	no	no	No

[c] significant at the 1% level, [a] significant at the 5% level, [b] significant at the 10% level
NB. East as in Tables 2 and 4, Small = Cyprus, Iceland and Wales

although only the country fixed effects were jointly significant. In addition to GDP per capita, three dummy variables were included: one for years when there was no World Cup or European Championship being played (these are the odd numbered years), one for host countries in championship years and a third for the winner in championship year.

The regression results are reported in Table 5. It is perhaps not surprising to find that the GDP variable is insignificant in all specifications given the descriptive statistics. However, more striking is the significance of World Cup and European Championship years. It was hypothesized that these coefficients would be *positive*, in the expectation that demand would stimulated by competitions taking place in the previous summer (e.g. 2003 refers to the season 2002/2003, immediately following the Japan/Korea World Cup). However, the coefficient is negative, suggesting that demand is higher in the season immediately preceding a major championship. This might reflect either a higher level interest in football in the run up to a championship (an expectations effect), or a lower level of interest due to saturation following a championship (substitution effect). This is a question that merits further investigation.

Hosting effects on demand are not significant, which is perhaps surprising given the amount of time and effort devoted to stimulating interest in football for these events. However, winners enjoy a substantial boost in demand in the season following their win. This effect was documented by Hirata and Szymanski (2009).

5 Conclusions

It would reasonable to expect that attendance at league football matches would be significantly affected by the economic cycle. There is certainly evidence that the current economic crisis has impacted on attendance in European countries.

However, the evidence also suggests that the link between GDP growth and attendance demand is at best tenuous. This is puzzling given that one would expect income effects to play a role in demand. Several explanations might explain these findings. First, the analysis has not controlled for the effect of team performance, which may affect demand and therefore the league aggregate in any given season. Second, it may be that income effects are offset by price movements, since most clubs exercise a degree of market power in relation to their fans, at least in the short run. Thus demand effects might appear in the revenues of clubs but not in attendance figures. Third, it may be mismatching due the timing of seasons and annual GDP measurement which mask underlying impacts, suggesting that a model based on quarterly data might be more effective.

The absence of cyclical effects does not imply that clubs are immune to the macroeconomic environment. Indeed, the finding that demand is sensitive to years in which major summer tournaments are held suggest a degree of substitution in consumption expenditures by fans. The difficulty from the perspective of the club is the apparent lack of predictability in the relationship between GDP growth and demand. Since clubs tend to survive on the edge of insolvency in most countries, spending everything they earn and barely breaking even, this means that unexpected variability in income can significantly destabilise the financial position of a club.

References

Borland, J., & R. Macdonald (2003). Demand for sport. *Oxford Review of Economic Policy, 19*(4), 478–502.

Dobson, S., & J. Goddard (1995). The demand for professional league football in England and Wales: 1925–1992. *The Statistician, 44*, 259–277.

Dobson, S., & J. Goddard (1996). The demand for football in the regions of England and Wales, *Regional Studies, 30*, 443–453.

Forrest, D., & Simmons, R. (2002). Team salaries and playing success in sports: A comparative perspective. *Zeitschrift für Betriebswirtschaft, 66*, 221–237.

Garcia-del-Barrio, P., & S. Szymanski (2009). Goal! Profit maximization vs win maximization in soccer leagues. *Review of Industrial Organization, 34*(1), 45–68.

Hall, S., Szymanski, S., & Zimbalist, A. (2002). Testing causality between team performance and payroll: The cases of major league baseball and English soccer. *Journal of Sports Economics, 3*(2), 149–168.

Hirata, T., & Szymanski, S. (2009). The J. league and the world cup. *Journal of the Japan Society of Sports Industry, 19*(1), 41–54.

Humphreys, B. (2002). Alternative measures of competitive balance in sports leagues. *Journal of Sports Economics, 3*(2), 133–148.

Schmidt, M., & Berri, D. (2001). Competitive balance and attendance: The case of major league baseball. *Journal of Sports Economics, 2*(2), 145–167.

Simmons, R. (1996). The demand for English league football: A club level analysis. *Applied Economics, 28*, 139–155.

Simmons, R. (2006). The demand for spectator sport. In W. Andreff & S. Szymanski (Eds.), *Handbook on the economics of sport*. Cheltenham: Edward Elgar.

Szymanski, S. (2003). The economic design of sporting contests. *Journal of Economic Literature, XLI*, 1137–1187.

Szymanski, S., & Smith, R. (1997). The English football industry, profit, performance and industrial structure. *International Review of Applied Economics, 11*(1), 135–153.

Economics of Gambling on Sports: A Multistage Stochastic Programming Approach to American Jai Alai Gambling Strategies

Qipeng P. Zheng, Yingyan Lou, and Panos M. Pardalos

Abstract With the increased gambling tolerance, the economic stake of gambling on sports is growing bigger than ever before. One common type is the para mutual gambling, where the gamblers bet on the results of games, such as Jai Alai, dog racing, horse racing, etc. Uncertainties lie both in the game itself and in the bets made by all gamblers. This study attempts to develop the optimum betting strategies for Jai Alai games based on the rules, the historical results and the random returns of each game. Two main concerns of this study are: modeling the randomness in both the game and the gambler sides, and formulating multistage stochastic mixed-integer models for the strategy-making problem to maximize the return and control the risk as well.

1 Introduction

Since the 1970s, sports have become more economically important than ever before. Gambling on sports is also flourishing due to the increased tolerance on gambling worldwide, and helps make the economic stake of sports even bigger, as is discussed in (Andreff & Szymański, 2006).The impact of gambling is even more vital to some specific sports, such as Jai Alai, horse and dog racing, etc. For example, in United States these sports would not even exist in those states where gambling is prohibited. On the other hand, in order to organize and manage sports more efficiently and increase its importance, operations research has been widely applied in sports management (Butenko, Gil-Lafuente, & Pardalos, 2004). Due to the advances of

Q.P. Zheng (✉)
Center for Applied Optimization, Department of Industrial & Systems Engineering, University of Florida, Gainesville, FL 32611, USA
e-mail: zqp@ufl.edu

S. Butenko et al. (eds.), *Optimal Strategies in Sports Economics and Management*, DOI 10.1007/978-3-642-13205-6_11,

optimization techniques, sports management models have been becoming more realistic and useful than ever. This study focuses on modeling and finding para mutual gambling strategies that can accommodate the trade-off between profit and risk based on a multistage stochastic programming approach. As a typical para mutual gambling game, betting strategies for Jai Alai are studied. This sport originated in Spain Basque area three centuries ago. Similar to the racket ball, this game is played between two opposing teams in an enclosed rectangular fronton. The players alternatively catch the ball and throw it to the front wall by using a big basket tied on one hand. There are two types of games: single and double, in which each team has one or two players respectively. In United States, there are usually eight teams in a Jai Alai game. All teams are arranged in a queue and each time the first two teams in the queue play against each other. The winner gets the point(s) and goes to the top of the queue, and the loser goes to the bottom without gaining any point. The rank of the eight teams is determined at the end of each game. Bets can only be placed before the game and the payoffs are calculated based on the wagering poll. Given the initial sequence of the eight teams in the queue and the payoffs, the betting strategies are investigated in this paper.

There are many ways to wager, such as "WIN", "PLACE", "QUINIELA", etc, and a good betting strategy should take into account the uncertainties lying in both the game results and the payoffs. The results of Jai Alai games were widely studied in the literature. Goodfriend & Friedman (1975, 1977) analyzed the results using Monte Carlo simulation. The winning probability of each team was examined in (Moser, 1982) using a tree formulation. Skiena (1988) further discussed the impact of limiting point on the results. More recently, Markov Chain analysis was applied to simulate the Jai Alai results in (Byrne & Hesse, 1996). On the other hand, the randomness in the payoffs is still not very clear. This is because that there is not enough data to compute the real expected return due to the large number of betting combinations and the fact that the composition of gamblers varies from time to time. Limited research has been done on Jai Alai betting strategies. Lane and Ziemba (2004) studied the Jai Alai arbitrage strategies and developed the perfect arbitrage conditions for the Mexican Jai Alai game. However, the Mexican gambling rules differ a lot from what are used in United States, thus their problem setting is quite different than the one studied in this paper.

In this study, the determination of a betting strategy is modeled as a multi-stage planning problem which considers uncertainties coming from both the game itself and the gamblers. Randomness in the game results is replicated using Monte Carlo simulation and the payoffs are assumed to follow a simple distribution. Data simulation is described in Sect. 2. Multistage stochastic betting strategy problems with different objectives are formulated in Sect. 3. Two other simple strategies are studied as well. Numerical examples and comparisons between different strategies are provided in Sect. 4 with discussion on risk arbitrage conditions. Section 5 concludes the paper.

2 Modeling Randomness in the Game

2.1 Game Results Simulation

We use Monte Carlo simulation to calculate the probability of each scenario. Similarly to Skiena (2001), we put eight teams in a queue and simulate the game between the top two teams. Instead of simulating the players' detailed actions (which is difficult to do and is not the main point here), the result of a particular game is calculated by applying a Bernoulli trial based on the winning probability (obtained from Jai Alai websites) of each team. Simulation of the match between the top two teams is repeated sequentially until a winner team (rank 1) can be determined. To find out the teams in the second and third place, playoff rules are then introduced. The playoff simulation is similar to the previous one, but a queue with the remaining seven teams is used instead. The playoff rules are made by law. Details can be found in (Skiena, 2001). To sum up, the simulation procedure is as follows:

Step 1. If the top team gets points equal to or higher than the winning points, go to step 3. Otherwise, go to step 2.

Step 2. Use Bernoulli trial to determine the winner and the loser for the game between the first two teams by using the previous records of these two teams. Winner gets two more points if all eight teams have played at least once, and gets one more otherwise. Winner goes to the top of the queue and loser goes to the end of the queue. All other teams move up by one position. Go to step 1.

Step 3. Playoff.

2.2 Payoffs

Payoffs will be affected by individual bets through their contribution to the wagering pool. This type of interaction is very difficult to model explicitly. To simplify the problem, we assume the payoffs are random variables independent to the gambler's betting decisions. This assumption can only be justified when the gambler's wager is small enough to be neglected compared to the whole wager pool. Therefore, the maximum amount of bet each gambler can place should be restricted. For modeling purposes, the payoffs are further assumed to follow a Bernoulli(p) distribution with $p = \frac{1}{2}$. That is to say, there are two equally likely outcomes of the payoffs, and we assume the payoffs are $\sigma_H b$ and $\sigma_L b$ where b is the amount of bet. However, if a realistic payoff distribution is to be modeled, the number of possible final scenarios would grow exponentially with the size of the finite sample drawn to approximate the actual distribution. Denote the payoff for a particular bet b by $\sigma(b, t, sp)$, where t denotes the stage and $sp \in \{1, 2\}$ denotes the payoff scenario, either high or low.

The average payoff data is obtained from (Skiena, 2001). The return of the bet for each game can then be calculated based on $\sigma(b, t, sp)$. The format of the data should be adjusted for different formulations. This will be explained in more details in Sect. 3.

3 Determination of Betting Strategies

3.1 Scenario-Based Multistage Stochastic Betting Strategy

Consider two types of wagering,

- "PLACE" — The gambler wins if his/her chosen team finishes first or second.
- "QUINIELA" — The gambler wins if his/her chosen two teams finish first and second in any order.

We first consider the tree representation of the problem. Figure 1 illustrates the tree structure of the problem. The nodes denoted by a circle are decision nodes where major betting decisions are made. Arcs in the tree represent the possible outcomes of the game results and payoff scenarios. Total returns of betting on the T Jai Alai games are calculated at the leaf (square) nodes. Note that as we are only interested in the combination of top two ranked teams without any exact order information, we only need to consider $\binom{8}{2} = 28$ possible game results. For each game result realization, there are two possible payoffs. Hence, the total number of arcs disseminating from each decision node is 56. The number of nodes thus grows exponentially on a base of 56 at each stage. More specifically, there are 56^{t-1} decision nodes at stage t and 56^T leaf nodes at the end of the gambling game.

Alternatively, the full horizon scenarios can be considered directly without specifying the history of the process. This way, nonanticipativity constraints (Birge & Louveaux, 1997) are needed to link related scenarios. Figure 2 presents the full-horizon representation of the problem. As in Fig. 1, circle nodes are decision nodes and square nodes are leaf nodes. In this representation, the total number of decision nodes is expanded to $T56^T$ instead of $(56^T - 1)/55$ in the previous one. The solid

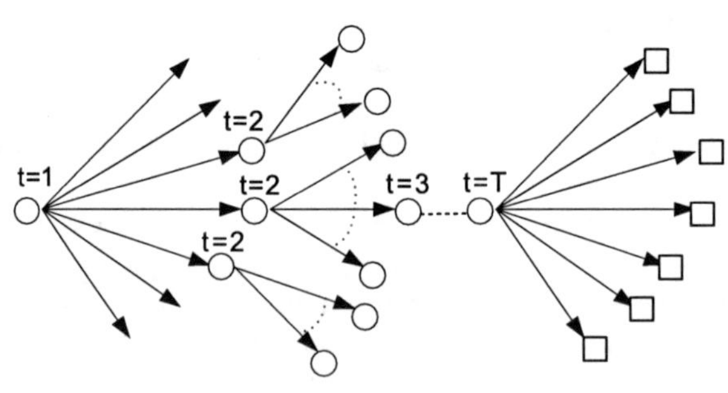

Fig. 1 Tree representation

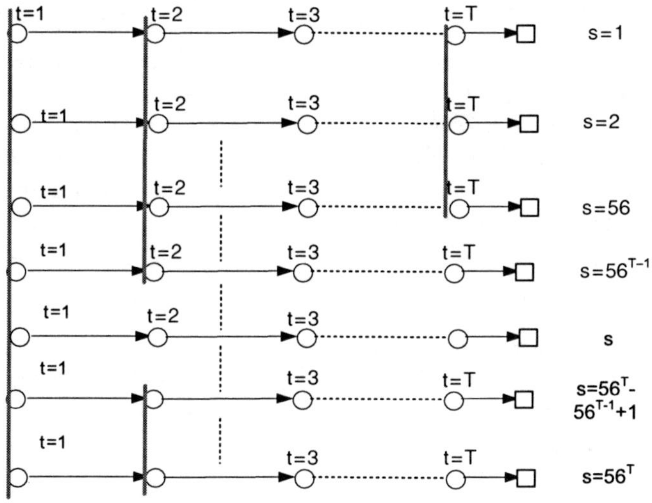

Fig. 2 Full horizon representation

vertical lines give an idea of the linkage between related scenarios, those who share common previous states. The full-horizon representation actually expands all the decision nodes in the tree representation, and the nonanticipativity constraints serve as the linkage to group them back.

3.1.1 Decision Variables

As discussed above, the bets placed by the gambler may affect the payoffs. To minimize this effect, we assume the gambler can only make at most one bet on each different combinations. Corresponding to "PLACE" and "QUINIELA" betting rules, two groups of major decision variables are introduced. For $i, j = 1, \ldots, 8$:

- x_i: binary variable denoting whether to place bet on player i for "PLACE" game;
- x_{ij}: binary variable denoting whether to place bet on players at positions (i, j) for "QUINIELA" game. (Here the order does not matter. Therefore, this group has 28 variables in all.)

These 36 binary decision variables compose the decision vector denoted as $x(b)$ at each decision node. Auxiliary variables y and z are introduced at each leaf node to measure the total gain and loss.

For the tree representation, major decision variables can be expressed as $x(b, t, k_t)$, where k_t is the index of the decision nodes at stage t. For the full-horizon representation, major decision variables are $x(b, t, s)$ where s denotes a full scenario. Since the number of the leaf nodes is the same as the number of total scenarios, we refer to the auxiliary variables as $y(s)$ and $z(s)$ in both representations.

3.1.2 Constraints

The constraints for this problem are relatively simple. The most important one is that at each decision node, the total bets made by the gambler cannot exceed the money available to him at that node. Another constraint is needed to calculate the net profit at the leaf nodes. For the full-horizon representation, a set of nonanticipativity constraints are further introduced. Other constraints include non-negativity of all the decision variables and binary requirement for the decision variable x.

3.1.3 Betting Strategies

After the above discussion on decision variables and constraints, the problem is now ready to be formulated as a mathematical program. Two different objectives, maximizing the expected profit and minimizing the conditional value-at-risk (CVaR), are studied in this section.

Profit Maximization

The most straightforward objective is to maximize the expected return. Under the tree representation, the problem can be formulated explicitly as follows,

[EP-T]

$$\text{Max} \sum_{s \in S} Prob(s)(y(s) - z(s)) \tag{1}$$

$$\text{s.t. } I(t, k_t) = \sum_{b=1}^{36} p(b)x(b, t, k_t), t = 1, \ldots, T, \forall k_t \in K_t, \tag{2}$$

$$R(t, k_t) = \sum_{b=1}^{36} \sum_{\xi=1}^{56} r(b, t-1, \xi)x(b, t-1, l_{k_t}),$$

$$t = 2, 3, \ldots, T, \forall k_t \in K_t, \tag{3}$$

$$w(1, k_1) = M, \tag{4}$$

$$w(t, k_t) = R(t-1, k_t) + w(t-1, l_{k_t}) - I(t-1, l_{k_t}),$$

$$t = 2, \ldots, T, \forall k_t \in K_t, \tag{5}$$

$$w(T+1, s) = \sum_{b=1}^{36} \sum_{\xi=1}^{56} [r(b, T, \xi)x(b, T, l_s)]$$

$$+ w(T, l_s) - I(T, l_s), s \in S \tag{6}$$

$$w(T+1, s) - M - y(s) + z(s) = 0, s \in S, \tag{7}$$

$$I(t, K_t) \leq w(t, K_t), t = 1, \ldots, T, \forall k_t \in K_t \tag{8}$$

$$y(s) \geq 0, z(s) \geq 0, \forall s \in S, \tag{9}$$

$$x(b, t, k_t) \in \{0, 1\}, t = 1, \ldots, T, b = 1, \ldots, 36, \forall k \in K_t. \tag{10}$$

In the above [EP-T] model, $p(b)$ is the ticket price of a bet b; K_t is the set of decision nodes at stage t; $r(b, t, \xi)$ is calculated through the payoff data, representing the return under game result/payoff scenario ξ for a bet b made at stage t; l_q denotes the branch node of node q in the tree and M is the initial amount of money the gambler has; $w(t, k_t)$ denotes how much money the gambler has at the beginning of game t. Constraints (2) and (3) define the investment, i.e., cash outflow, and the total return at each decision node respectively; constraint (4), (5), and (6) calculate/define the total available money for each node (both decision and leaf nodes); constraint (7) then computes the total gain and loss for each leaf node; finally constraint (8) restricts the investment at each decision node to be no more than the available money at that node. If the total available money is zero or less than the minimum ticket price, then the gambler will stop gambling. Note that $w(t, k_t)$ and $w(T + 1, s)$ are always non-negative, as there is no penalty to the unsuccessful bets, and the gambler only can bet what he has.

Alternatively, under the full-horizon representation, the problem can be modeled as follows,

[**EP-F**]

$$\text{Max} \sum_{s\in S} Prob(s)\left[y(s) - z(s)\right] \tag{11}$$

$$\text{s.t. } I(t, s) = \sum_{b=1}^{36} p(b)x(b, t, s), t = 1, \ldots, T, \forall s \in S, \tag{12}$$

$$R(t, s) = \sum_{b=1}^{36} r(b, t, s)x(b, t, s), t = 1, \ldots, T, \forall s \in S, \tag{13}$$

$$\begin{aligned} w(t, s) &= R(t-1, s) + w(t-1, s) - I(t-1, s),\\ t &= 1, \ldots, T+1, \forall s \in S, \end{aligned} \tag{14}$$

$$w(T + 1, s) - M - y(s) + z(s) = 0, \forall s \in S, \tag{15}$$

$$w(1, s) = M, \forall s \in S, \tag{16}$$

$$I(t, s) \leq w(t, s), t = 1, \ldots, T, \forall s \in S, \tag{17}$$

$$\sum_{\gamma\in S^t_{J(s,t)}} Prob(\gamma)\left[x(b, t, \gamma) - x(b, t, s)\right] = 0, t = 1, \ldots, T, \quad b = 1, \ldots, 36, \forall s \in S, \tag{18}$$

$$(9) - (10).$$

In the above [EP-F] model, constraints (12), (13), (14), (15), (16), and (17) are almost the same as in [EP-T], except that they are defined for each node in the full-horizon representation (see Fig. 2). Note that $r(b, t, s)$ is not the same parameter as $r(b, t, \xi)$, though they both represent returns. The expanded parameter $r(b, t, s)$ is computed by correctly assigning $r(b, t, \xi)$ to each full-horizon scenario.

Equation (18) is the nonanticipativity constraint, where $S^t_{J(s,t)}$ denotes the set of scenarios that share the common history of s. This constraint enforces the knowledge of future outcomes to be excluded when the decision at current stage is made.

Although the full-horizon representation is relatively simple in terms of modeling efforts, it has a major drawback: the space and time complexity of solving the problem increases exponentially. This is because the expansion of all the decision nodes in the tree representation increases both the number of decision variables and the number of constraints. The nonanticipativity condition further adds $B \times T \times |S|$ constraints, where B is the total number of decision variables at each node. Therefore, it requires a lot memory to store the coefficient matrices. More importantly, as a mixed-integer program, the computational time and the memory required by the Branch-and-Bound (e.g., Floudas (1995)) procedure grow exponentially with the size of the problem. As demonstrated later in the paper, standard solvers are not able to handle this formulation due to lack of memory.

Optimization Using CVaR

Besides the expected profit, a gambler may also be concerned about the average loss in high consequence scenarios. Rockafellar & Uryasev (2000) proposed a linear formulation to calculate value-at-risk (VaR) and optimize CVaR simultaneously. This is done by introducing a free variable ζ and non-negative variables $\eta(s)$ for each leaf node. Intuitively, $\eta(s)$ measures the loss exceeding a certain value ζ, and the sum of the expected excess loss and ζ provides a measure of the average loss in high consequence scenarios. Following this idea, the betting strategy considering both the expected profit and the average excess losses can be formulated, based on the tree representation, as:

[CVaR-T]

$$\text{Min} \quad Z = \zeta + (1-\beta)^{-1} \sum_{s \in S} Prob(s)\eta(s) - \lambda \sum_{s \in S} Prob(s)\left[y(s) - z(s)\right] \quad (19)$$

$$\text{s.t.} \quad (2), (3), (4), (5), (6), (7), (8), (9), \text{ and } (10),$$

$$\eta(s) \geq z(s) - y(s) - \zeta, \forall s \in S, \quad (20)$$

$$\eta(s) \geq 0, \forall s \in S, \quad (21)$$

where β is a percentage parameter. Since there is generally a trade-off between expected profits and the risks (represented here by CVaR), λ in Eq. (19) is a weighing factor representing the gambler's attitude toward risk. The smaller the λ is, the more risk-adverse the gambler is. Note that because gamblers are more concerned with losses, it is the right tail of the distribution of $z(s) - y(s)$ that we should model.

3.2 Dynamic Programming Formulation and Optimal Policies

Since the introduction of dynamic programming by Bellman in 1952, it has become a very important and popular modeling and solution tool for a great variety of operations research problems. Many dynamic approaches to solve sports management problems have been proposed and conducted as well, such as (Sackrowitz, 2004). Dynamic programming is a very useful tool to solve multistage deterministic/stochastic problems (Bertsekas, 1987). Because the results of each stage are independent of each other and the current stage is affected only by its immediate predecessor, using multistage dynamic programming can both capture the dynamics across all the stages and reduce the size of the problem since there is no need to formulate all the outcomes if the future cost/benefit function can be modeled appropriately. If the dimensionality explosion can be handled properly, we can include all possible decisions, e.g., all types of bets, instead of just "PLACE" and "QUINIELA". Let $x^t_{k,l}$ denote the amount of bets placed on combination l for the bet of type k at stage t. It is an integer variable, since in most of Jai Alai places the bets are discrete. For example, the minimum bet is one dollar for any of your favorite combination. In this section, this constraint is relaxed and every variable is considered to be continuous. Suppose we only want to maximize the ultimate value, i.e., the money you will have after the last game. So there will be no intermediate objective function. The multistage dynamic programming formulation of the profit maximization problem at stage t ($t < T$) for scenario j of stage $t-1$ is as follows,

[DP$_t$]

$$
\begin{aligned}
B_{t-1}\left(x^{t-1}, w^{t-1}, \xi_j^{t-1}\right) = \text{Max} \quad & B_t(x^t, w^t) \\
\text{s.t.} \quad & w^t + \sum_{k\in K}\sum_{l\in L_k} x_l^t = w^{t-1} + \sum_{k\in K}\sum_{l\in L_k} R_{j,l}^{t-1} x_l^{t-1}, \\
& x_l^t \geq 0, \forall l \in L_k, k \in K, \\
& w^t \geq 0,
\end{aligned}
$$

where $R^t_{j,l}$ is the return when combination l wins at stage t under scenario j; w^t is the money the gambler does not bet at stage t; ξ_j^{t-1} is the outcome of scenario j of stage $t-1$; $B_t(x^t, w^t)$ is the future benefit function which is defined as follows,

$$
B_t(x^t, w^t) = \sum_{s\in S} P_s^t B_t(x^t, w^t, \xi_s^t)
$$

where P_s^t denotes the probability of scenario s of stage t. The last stage problem, [DP$_T$], which does not have a future benefit function, is shown as follows,

[DP$_T$]

$$\begin{aligned} B_{T-1}\left(x^{T-1}, w^{T-1}, \xi_j^{T-1}\right) = \text{Max} \quad & w^T + \sum_{k\in K}\sum_{l\in L_k}\sum_{s\in S} P_s^T R_{s,l}^T x_l^T \\ \text{s.t.} \quad & w^T + \sum_{k\in K}\sum_{l\in L_k} x_l^T \\ & = w^{T-1} + \sum_{k\in K}\sum_{l\in L_k} R_{j,l}^{T-1} x_l^{T-1}, \\ & x_l^T \geq 0, \forall l \in L_k, k \in K, \\ & w^T \geq 0. \end{aligned}$$

The solution of the problem [DP$_T$] is not hard to find,

$$B_{T-1}\left(x^{T-1}, w^{T-1}, \xi_j^{T-1}\right) = \begin{cases} PR_{\max}^T \left(R_{j,l}^{T-1} x_l^{T-1} + w^{T-1}\right) & PR_{\max}^T > 1 \\ R_{j,l}^{T-1} x_l^{T-1} + w^{T-1} & o/w \end{cases}$$

where $PR_{\max}^T = \text{Max}\left\{P_l^T R_{s,l}^T | s \in S, l \in L_k, k \in K\right\}$.

Let $ER_{\max}^T = \text{Max}\left(PR_{\max}^T, 1\right)$, and then we can rewrite the objective function of stage $T-1$ as follows,

$$B_{T-1}\left(x^{T-1}, w^{T-1}, \xi_j^{T-1}\right) = ER_{\max}^T \left(w^{T-1} + \sum_{k\in K}\sum_{l\in L_k}\sum_{s\in S} P_s^{T-1} R_{s,l}^{T-1} x_l^{T-1}\right). \tag{22}$$

The stage $T-1$ problem is actually the same as in stage T, since $ER_{\max}^T$ is a constant in stage $T-1$. The best policy is still to choose the combination with the highest $ER_{\max}^{T-1}$. So do all the previous stages. Hence for a relaxed problem, the optimal policy for stage t is to choose the combination with the highest $ER_{\max}^t$, and then bet all your money on this combination if $ER_{\max}^T > 1$ and do not wager otherwise.

This policy in long run is the best, but it may be very dangerous in short run, since the series of games will never repeat and the formulation does not include any risk constraint. In order to control risk, we can add CVaR constraints in the above dynamic formulation since they are linear constraints and easy to handle. The formulation is as follows,

[DP-CVaR$_T$]

$$\begin{aligned} B_{T-1}\left(x^{T-1}, w^{T-1}, \xi_j^{T-1}\right) = \text{Max} w^T & + \sum_{k\in K}\sum_{l\in L_k}\sum_{s\in S} P_s^T R_{s,l}^T x_l^T \\ \text{s.t.} \quad w^T + \sum_{k\in K}\sum_{l\in L_k} x_l^T & = w^{T-1} + \sum_{k\in K}\sum_{l\in L_k} R_{j,l}^{T-1} x_l^{T-1}, \end{aligned}$$

$$\sum_{k\in K}\sum_{l\in L_k} x_l^T - \sum_{k\in K}\sum_{l\in L_k} R_{\gamma,l}^{T-1} x_l^{T-1} \leq \eta + y\gamma^T, \forall \gamma \in S,$$

$$\eta + \sum_{\gamma\in S} \frac{P_\gamma^T y\gamma^T}{1-\alpha} \leq \psi,$$

$$y\gamma^T \geq 0, \forall \gamma \in S,$$

$$x_l^T \geq 0, \forall l \in L_k, k \in K,$$

$$w^T \geq 0.$$

It is not that obvious to find the solution of this sub problem because of the risk constraints. But it is still easy to find a sub optimal solution, the one which has the highest expected return, $ER_{\max}^t$, great than 1 and whose conditional value at risk is lower than ψ. A sub optimal gambling policy would be to bet all money on a combination which satisfy the above conditions, and bet nothing if no such a combination exist.

3.3 Other Betting Strategies

3.3.1 Proportion-Based Formulation

One possible simple strategy is that we only bet a certain percent, α, of the money we have at each stage on the combination with the maximal expected net gain. With this strategy, the betting decision (which combination to bet on) is predetermined. Therefore, the scenarios of the game can be simplified to two: whether this combination wins or not. Figure 3 is a demonstration for a simple two-stage game. Assume the winning probability of the combination a gambler bet on is $Prob(t)$ for stage t, and the average return is R_t. At each node (t, K_t), denote the available

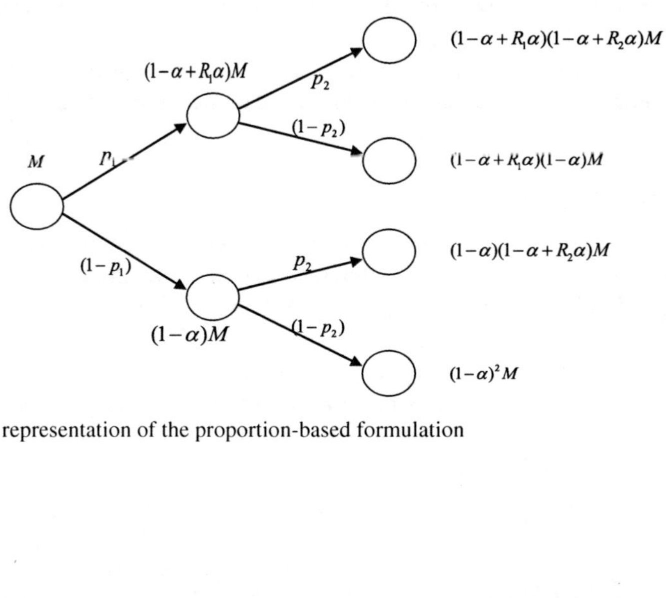

Fig. 3 Tree representation of the proportion-based formulation

money on hand by $m_{t,k}$, and $p_{t,k}$, the probability this node can be reached. For each node in the simplified scenario tree, the available money and the probabilities can be calculated. Thus, maximizing the expected return with respect to α in a T stage problem is formulated as:

[PB]

$$
\begin{aligned}
\text{Max} \ & \sum_{j=1}^{2^T} p_{T,j} m_{T,j} \\
\text{s.t.} \ & Prob(i+1) p_{i,j} = p_{i+1,2j-1}, j = 1, \ldots, 2^i, i = 0, \ldots, T-1, \\
& (1 - Prob(i+1))\, p_{i,j} = p_{i+1,2j}, j = 1, \ldots, 2^i, i = 0, \ldots, T-1, \\
& (1-\alpha) m_{i,j} + \alpha m_{i,j} R_i = m_{i+1,2j-1}, j = 1, \ldots, 2^i, i = 0, \ldots, T-1, \\
& (1-\alpha) m_{i,j} = m_{i+1,2j}, j = 1, \ldots, 2^i, i = 0, \ldots, T-1, \\
& m_{0,1} = M, \\
& 0 \le \alpha \le 1, \\
& m_{i,j} \ge 0, j = 1, \ldots, 2^i, i = 0, \ldots, T-1.
\end{aligned}
$$

However, the result of this optimization model may not be a good strategy for the Jai Alai case, since the returns are greatly affected by the distribution of wagers made by the gamblers. If a gambler always wagers a certain proportion of his available money, the assumption that the payoffs are independent to the gambler's decision may not be valid anymore. Suppose the money bet on this combination by all the gamblers is n, and the empirical average return per dollar (without the influence of the gambler's bet) is R. If a gambler bets m dollars on this combination and he wins, then the real average return per dollar, $R^*(m)$, is computed as $R^*(m) = \frac{nR}{m+n}$. This is a decreasing function of m when it is nonnegative. Since the wager pool for one particular combination is not very large, $R^*(m)$ is always no greater than R and very sensitive to m. Because n is hard to be observed (not released by the Jai Alai frontons), it is difficult to predict the real average return in the objective function.

3.3.2 Simple Arbitrage Strategy

Suppose that all Jai Alai games are independent of each other, and our bets do not affect the returns. It is not difficult to perceive the arbitrage strategy: make bets on the team/combination k, which satisfies the following property, $p_k \times R_k > m_k$, where p_k is the winning probability, R_k is the average return, and m_k is the money we bet on this combination. This strategy is used in (Skiena, 2001). In the long run (infinite stages), this strategy leads to no loss of money. This observation is pretty rough and sensitive to the assumptions made.

4 Numerical Examples

For demonstration, we consider a toy problem with only two stages. A gambler initially has ten dollars to place bets in either "PLACE" or "QUINIELA" games. Eight players are simulated for each Jai Alai game to obtain the winning probabilities. Table 1 reports the simulated probability of a combination i, j appearing both in the first two places for the two games. Average payoff data in (Skiena, 2001) for "PLACE" and "QUINIELA" games are summarized in Tables 2 and 3 respectively. All tickets are worth two dollars. The models are implemented in the general algebraic modeling system GAMS (Brook, Kendirck, & Meeraus, 1992) and the commercially available solver CPLEX is used to solve these mixed-integer programs.

We first solve the expected profit maximization problem. For the [EP-F] formulation, an "out of memory" error is thrown and GAMS is not able to generate the model with 1,888MB memory. For the [EP-T] formulation, the model is solved within several minutes. The optimum objective value is 16.45 and the detailed betting strategy is reported in Table 4. According to the solution, the gambler should

Table 1 Winning probabilities for the two games

	Winning Probability	
Player combination	Game1	Game2
1,2	0.10348	0.06257
1,3	0.10618	0.06343
1,4	0.0122	0.05701
1,5	0.15607	0.04696
1,6	0.03913	0.03993
1,7	0.04196	0.0343
1,8	0.02933	0.03453
2,3	0.06986	0.06321
2,4	0.00793	0.05756
2,5	0.10861	0.04735
2,6	0.02687	0.04017
2,7	0.02667	0.03407
2,8	0.01777	0.03451
3,4	0.00453	0.03796
3,5	0.09345	0.03601
3,6	0.02137	0.0352
3,7	0.02562	0.03344
3,8	0.01794	0.03744
4,5	0.01391	0.02183
4,6	0.00392	0.02468
4,7	0.00446	0.02832
4,8	0.00261	0.03328
5,6	0.01626	0.01351
5,7	0.02148	0.0194
5,8	0.01712	0.0261
6,7	0.00487	0.0113
6,8	0.00428	0.01665
7,8	0.00217	0.0093

Table 2 "PLACE" payoffs for the two games

Player	Payoffs	
	Game1	Game2
1	$5.91	$6.17
2	$5.80	$6.01
3	$6.54	$6.76
4	$7.06	$7.77
5	$7.51	$7.49
6	$8.51	$8.71
7	$8.25	$8.35
8	$8.61	$8.89

Table 3 "QUINIELA" payoffs for the two games

Player combination	Payoffs	
	Game1	Game2
1,2	$47.21	$37.55
1,3	$39.99	$40.33
1,4	$41.08	$45.05
1,5	$46.25	$38.44
1,6	$51.70	$45.77
1,7	$50.56	$46.23
1,8	$55.07	$57.22
2,3	$35.85	$35.96
2,4	$39.64	$34.11
2,5	$44.41	$36.39
2,6	$61.83	$45.45
2,7	$44.17	$45.49
2,8	$57.06	$53.31
3,4	$51.46	$49.30
3,5	$43.35	$50.13
3,6	$54.32	$47.52
3,7	$48.38	$58.13
3,8	$47.23	$58.21
4,5	$66.51	$65.65
4,6	$62.83	$61.02
4,7	$46.99	$49.13
4,8	$61.85	$53.89
5,6	$84.20	$97.31
5,7	$55.56	$52.81
5,8	$65.37	$56.55
6,7	$88.73	$96.83
6,8	$71.07	$70.37
7,8	$90.40	$82.51

bet on more risky "QUINIELA" combinations in the first stage, and the more conservative "PLACE" combinations in the second stage. The solution to the CVaR-T model with $\beta = 0.9$ and $\lambda = 0.65$ is summarized in Table 5. With this strategy, the expected return is 5.55, less than that of the [EP-T] model; but the 90%-CVaR is 2 (while that of [EP-T] solution is 10), indicating the gambler will lose only two dollars on average in the highest 10% loss scenarios. By adjusting the value of λ, the relationship between the risk measure and the expected return under the

Table 4 Results for [EP-T] Model

	Decision Node (t, s)										
Player combination	1,1	2,1	2,2	2,3	2,4	2,7	2,8	2,19	2,20	2,29	2,30
1		1	1	1	1	1	1	1	1	1	1
2		1	1	1	1	1	1	1	1	1	1
3		1	1	1	1	1	1	1	1	1	1
4		1	1	1	1	1	1	1	1	1	1
1,2	1	1	1	1	1	1	1	1	1	1	1
1,3	1	1	1	1	1	1	1	1	1	1	1
1,4		1	1	1	1	1	1	1	1	1	1
1,5	1										
2,3		1	1	1	1	1	1	1	1	1	1
2,5	1										
3,5	1										
3,8		1	1	1	1	1	1	1	1	1	1

Table 5 Results for [CVaR-T] Model with $\beta = 0.9$ and $\lambda = 0.65$

	Decision Node (t, s)				
Player combination	1,1	2,1	2,2	2,7	2,8
1		1	1	1	1
2		1	1	1	1
3		1	1	1	1
4		1	1	1	1
1,2	1	1	1	1	1
1,3		1	1	1	1
1,4		1	1	1	1
1,5	1				
2,3		1	1	1	1
3,8		1	1	1	1

optimal solution to [CVaR-T] model is plotted in Fig. 4. This observation confirms the trade-off between the two performance measures.

The [PB] model is solved as well with the same numerical settings. The proportion based strategy only bets on the combination with the maximal expected net gain. For the first game, the combination is "QUINIELA" (1, 5) with expected net gain of 2.61 per dollar, and "QUINIELA" (1, 4) with expected net gain of 0.28 per dollar is the one to bet on for the second game. The problem becomes,

$$\text{Max}\{M(0.74\alpha^2 + 2.89\alpha + 1)|0 \leq \alpha \leq 1\}$$

The optimal objective value is $4.63M$ with $\alpha^* = 1$. Whether this net gain is greater than the stochastic formulation solution depends on M. If M is big, the strategy leads to a larger expected profit. However, since not many people bet on "QUINIELA" game in practice (small value of n), this solution is usually an overestimate of the actual situation. If M is small, the expected profit will be lower than the stochastic optimization solution.

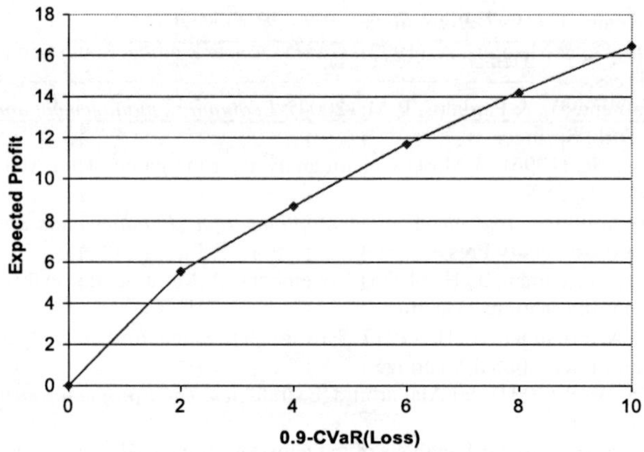

Fig. 4 Trade-off between expected profit and risk measure

As for the simple arbitrage strategy, consider the same two stage problem. Since the $p_k R_k > m_k$ condition holds for $k = 6, 7, 8$ at the first stage, rather than the strategy provided by the stochastic program, a gambler will make some bets on "PLACE" if this strategy is adopted. Given $m_k = 1$, the optimal expected profit using this strategy is only 6.01, less than the solution of the stochastic models.

5 Conclusion

In this study, optimal betting strategy problems for the Jai Alai games are formulated as mixed-integer scenario-based multistage stochastic programs. Several formulations to maximize the expected net gain or to control the risk are investigated. Two other straightforward strategies, the proportion-based and the simple arbitrage rules, are studied and treated as bench marks for the stochastic formulations. In our numerical experiments, the stochastic multistage planning formulation outperforms the others in terms of the expected net gain. Based on the solution of the stochastic programs, it seems that there might exist arbitrage in Jai Alai games. Future research will focus on the conditions under which there exists an arbitrage strategy. Also the computational issue is an interesting problem in the future, since the model will grow exponentially if all possible wagering combinations and more stages are considered. Furthermore, how to handle the sub problem of [DP-CVaR$_t$] to get an appropriate future benefit function is another problem of interest.

References

Andreff, W., & Szymański, S. (2006). *Handbook on the economics of sport*. Cheltenham: Edward Elgar.

Bertsekas, D. P. (1987). *Dynamic programming: Deterministic and stochastic models*. Englewood Cliffs, NJ: Prentic-Hall, Inc.

Birge, J. R., & Louveaux, F. (1997). *Introduction to stochastic programming*. New York: Springer.
Brook, A., Kendirck, D., & Meeraus, A. (1992). *GAMS: A user's guide*. South San Francisco, CA: The Scientific Press.
Butenko, S., Gil-Lafuente, J., & Pardalos, P. M. (2004). *Economics, management and optimization in sports*. New York: Springer.
Byrne, P. J., & Hesse, R. (1996). A Markov chain analysis of the game of Jai Alai. *Mathematics Magazine, 69*(4), 279–283.
Floudas, C. (1995) *Nonlinear and mixed integer optimization: Fundamentals and application*. New York: Oxford University Press.
Goodfriend, M. J., & Friedman, J. H. (1975, November). A Monte-carlo analysis of Jai Alai. ORSA/TIMS Joint International Meeting.
Goodfriend, M. J., & Friedman, J. H. (1977, November). Some further analysis of Jai Alai. ORSA/TIMS Joint International Meeting.
Lane, D., & Ziemba, W. T. (2004). Jai Alai arbitrage strategies. *The European Journal of Finance, 10*, 353–369.
Moser, L. E. (1982). A mathematical analysis of the game of Jai Alai. *The American Mathematical Monthly, 89*(5), 292–300.
Rockafellar, R. T., & Uryasev, S. (2000). Optimization of conditional value-at-risk. *Journal of Risk, 2*(3), 21–41.
Sackrowitz, H. (2004). Dynamic programming and time related strategies in sports. In S. Butenko, J. Gil-Lafuente, & P. M. Pardalos (Eds.), *Economics, management and optimization in sports*. New York: Springer.
Skiena, S. S. (1988). A fairer scoring system for Jai-Alai. *Interface, 33*(6), 35–41.
Skiena, S. S. (2001). *Calculated bets: Computers, gambling, and mathematical modeling to win*. New York: Cambridge University Press.

Lightning Source UK Ltd.
Milton Keynes UK
24 September 2010

160327UK00007B/23/P

9 783642 132049